THE SEVEN SISTERS
OF SLEEP

MORDECAI CUBITT COOKE

THE SEVEN
SISTERS
OF SLEEP

THE CELEBRATED DRUG CLASSIC

MORDECAI COOKE

Park Street Press
Rochester, Vermont

Park Street Press
One Park Street
Rochester, Vermont 05767
www.gotoit.com

Library of Congress Cataloging-in-Publication Data

Cooke, M. C. (Mordecai Cubitt), b. 1825.
    The seven sisters of sleep : the celebrated drug classic / Mordecai Cooke.
      p.   cm.
    Originally published: London : James Blackwood, 1860. With original subtitle: Popular history of the seven prevailing narcotics of the world.
    ISBN 0-89281-748-8 (alk. paper)
    1. Narcotics. 2. Psychotropic drugs. 3. Psychotropic plants. 4. Drug abuse. 5. Tobacco habit. I. Title. II. Title: Popular history of the seven prevailing narcotics of the world.
GT3010.C66 1997
394.1'4—dc21
                                        97-22404
                                        CIP

Printed and bound in the United States

10 9 8 7 6 5 4 3 2 1

Text design and layout by Kristin Camp
This book was typeset in Goudy and Willow

Frontispiece photograph included courtesy of the Hunt Institute for Botanical Documentation, Carnegie Mellon University, Pittsburgh, Pennsylvania.

Park Street Press is a division of Inner Traditions International

Distributed to the book trade in Canada by Publishers Group West (PGW), Toronto, Ontario
Distributed to the book trade in the United Kingdom by Deep Books, London
Distributed to the book trade in Australia by Millennium Books, Newtown, N. S. W.
Distributed to the book trade in New Zealand by Tandem Press, Auckland
Distributed to the book trade in South Africa by Alternative Books, Ferndale

CONTENTS

FOREWORD

Mordecai Cubitt Cooke wrote *The Seven Sisters of Sleep* in England in 1860, and both the place and time are essential to understanding the importance of this work—in its own era and to us today. Nineteenth-century Europe was an age, much like our own, when the respectable middle class was developing a healthy interest in the use of mind-altering drugs of all kinds. In Cooke's day this middle class itself had just developed. The democracies in France and the United States were still in their youth, the industrial revolution was stirring to life, and a class of people newly wealthy and educated, and curious about the world, were forming the first large market for mainstream books.

The Seven Sisters of Sleep was made to order for this audience. Cooke, a mycologist and director of the Metropolitan Scholastic Museum, understood that the scientific texts of the time were too dry and obscure for this readership, so without leaving the soggy comforts of England he set out to condense the available information into an entertaining text accessible to all. And how admirably he succeeded! Scholars Richard Evans Schultes and Michael Aldritch have even proposed that Cooke was Lewis Carroll's primary source for the psychedelic episodes in *Alice*

in Wonderland. What distinguishes his book above all from other popular drug books is the suppleness of his prose, the delightful humor he
brings to his subject, and his infectious curiosity for all the odd habits
of *Homo sapiens* of all colors and creeds.

Indeed, one of Cooke's notable qualities is his lack of prejudice
toward non-Europeans, so rare in the Victorian era. Cooke lived during the heyday of British colonialism, when many others wrote off the
rampant use of cannabis, opium, and coca by other cultures as weak-
nesses of their inferior characters. But of the seven "narcotics" Cooke
surveys—tobacco, opium, cannabis, betel nut, coca, datura, and fly
agaric—Europeans were addicted to the first as surely as the Chinese
were to the second, and Cooke is refreshingly clear-eyed in saying so:

> If we smoke our pipes of tobacco ourselves, while in the midst of
> the clouds, we cannot forbear expressing our astonishment at the
> Chinese and others who indulge in opium. Pity them we may, per-
> haps, looking upon them as miserable wretches the while, but they
> do not obtain our sympathies. Philanthropists at crowded assem-
> blies denounce, in no measured terms, "the iniquities of the opium
> trade," and then go home to their pipe or cigar, thinking them per-
> fectly legitimate, whether the product of slave labor or free. It is the
> same sort of feeling that the Hashasheens of the East inspire, and
> indeed all, who have a predilection for other narcotics than those
> which Johnny Englishman delights in, come in for a share of his
> contempt.

Sentiments such as these, pitched to a new class that—after centuries
of being told what to believe by the church and the monarchy—was
both ignorant of and hungry for the scientific truths of the world to a
degree not seen since, must have been quite controversial, and the
publication of this book can be seen as a supremely egalitarian act.
Perhaps too much so, for shortly after its publication *The Seven Sisters
of Sleep* disappeared. Cooke himself seems to have wanted it this way.
He settled into a comfortable life as a highly respected scientist and
never wrote about such controversial subjects again. For more than a
century the book has existed more as rumor than as reality.

Kudos, therefore, to publisher Ehud C. Sperling and Park Street
Press for their foresight in returning to a wide audience a book we
have much to learn from. In addition to the fascinating facts con-

tained herein concerning the use of drugs in the nineteenth-century world, *The Seven Sisters of Sleep* is a model for those who wish to consider the subject of drug use while staying above the swirling political winds of the day. This is the second in a trilogy of classic works on mind-altering plants that Park Street Press will publish, with Baron Ernst von Bibra's *Plant Intoxicants* published in 1995 and Louis Lewin's *Phantastica* promised for 1998. The debate over the use and abuse of drugs is sure to intensify in the years to come, and these fresh and insightful voices from the past will greatly help to keep us honest as we decide what shape our society will take.

July 1997
Rowan Robinson
Xinxiang Province, China

DEDICATION

To all lovers of tobacco, in all parts of the world,
juvenile and senile, masculine and feminine;
and to all abstainers,
voluntary and involuntary—
To all opiophagi, at home and abroad,
whether experiencing the pleasures, or pains
of the seductive drug—
To all haschischans, east and west,
in whatever form they choose
to woo the spirit of dreams—
To all buyeros, Malayan or Chinese,
whether their siri-boxes are full, or empty—
To all coqueros, white or swarthy,
from the base to the summit
of the mighty cordilleras—
To all votaries of stramonium and henbane,
highlander, or lowlander—
And to all swallowers of amanita,
either in Siberia or elsewhere—
these pages come greeting
with the best wishes
of their obedient servant.

The Author

PREFATORY PREMONITION

A certain miller was much annoyed by a goblin, who used to come and set his mill at work at night when there was no grain to be ground, greatly to the danger of the machinery, so he desired a person to watch. This person, however, always fell asleep, but once woke up from a nap time enough to see the mill in full operation, a blazing fire, and the goblin himself, a huge hairy being, sitting by the side thereof. "Fat's yer name?" said the Highlander. "Ourisk," said the unwelcome guest; "and what is yours?" "Myself," was the reply; "her nainsell." The goblin now went quietly to sleep, and the Highlander, taking a shovel of hot coals, flung them into the hairy lap of the goblin, who was instantly in a blaze. Out ran the monster to his companions, making as much noise as he could. "Well," said they, "who set you on fire?" "Myself," said the unlucky monster. "Well, then, you must put it out yourself," was the consoling rejoinder.

Some of my readers may arrive at the conclusion, that I, like the Ourisk, have trespassed upon other people's property, and ground my corn at their mill. Let it not be assumed, on my account, inasmuch as I do not myself make that assumption, that I have journeyed from Cornhill to Cathay, in search of those who habituate themselves to the indulgences herein set forth. Others have laboured, and I have eaten of the fruits of their labours. Travellers numberless have contributed to furnish my table, in some instances, without even thanks for their pains. This is the way of the world, and I am not a whit better than my neighbors. Let it, therefore, be understood, that I make no pretensions to aught beyond the form in which

these numerous contributions are now presented to the reader. The
tedium of wading through volume after volume in search of information on these subjects has been performed for him, and compacted
together into a pocket companion, saving, thereby, to him, a large
amount of trouble, and a small amount of vexation. Private corre-
spondence has furnished a portion of the information. Those who may
recognise my own poaching pranks upon their domains may throw
coals of fire upon my lap, and leave "Myself" to extinguish the flame.

Herein the reader will find only a popular history of the most
important narcotics indulged in, and the customs connected with that
indulgence. Mere statistical details have as much as possible been
avoided, and those calculated to interest the more matter-of-fact reader
added in a tabulated form, as an appendix. The majority of these tables
have been compiled from official documents, trade circulars, or com-
mercial returns, and care has been taken to render them correct up to
the period of their dates. In this department I am largely indebted to
the valuable assistance of P. L. Simmonds, Esq., F.S.S., to whom I thus
tender my thanks.

Those who are desirous of seeing specimens of the narcotics named
in the following pages, can visit either the Museum of the Royal
Botanic Gardens at Kew, the East India House Museum, the Food
Department in the gallery of the South Kensington Museum, or the
Industrial Museum in the gallery of the central transept of the Crystal
Palace, in each of which they will meet with some of the articles named,
though in none of them will they discover all. In the former two are
illustrations of the opium manufacture, and at Kensington an interesting series of tobaccos, and other articles connected with the indulgence therein, and also with opium-smoking in China, together with
some of the tobacco substitutes and sophistications. None of these
collections are so complete as they might be. Public museums of this
kind have every facility for doing more to instruct the public on the
common things of every-day life: why they do not accomplish this, is
as much a fault, perhaps, of the public as of themselves. There are
hopes, however, to be entertained that one, at least, of these institu-
tions will exhibit, in a complete and collected form, the principal
narcotics and their substitutes.

Why I should have chosen such a title for my volume, and wherefore invested it with a legend, is a matter of little importance. It was a

fancy of my own, and if any think fit to quarrel with it, they may do so, without disturbing my peace of mind. The reply of the Ourisk to his companions, as to who set him on fire, was, "Myself."

Parents seldom baptize their children with a name pleasing to all their friends and relatives, yet the child manages to get through the world with it, and—dies at last.

M. C. C.
Lambeth

CHAPTER I

SOMEWHAT FABULOUS

Oh sleep! It is a gentle thing,
Beloved from pole to pole.

Coleridge

During the Decian persecution, seven inhab-
itants of Ephesus retired to a cave, six were persons of some consequence, the seventh was their servnt; from hence they despatched the
attendant occasionally to purchase food for them. Decius, who like
most tyrants possessed long ears, hearing of this, ordered the mouth of
the cave to be stopped up while the fugitives were sleeping. After a
lapse of some hundred years, a part of the masonry at the mouth of the
cave falling, the light flowing in awakened them. Thinking, as Rip
Van Winkle also thought, that they had enjoyed a good night's rest,
they despatched their servant to buy provisions. All appeared to him
strange in Ephesus; and a whimsical dialogue took place, the citizens
accusing him of having found a hidden treasure, he persisting that he
offered the current coin of the realm. At length, the attention of the
emperor was excited, and he went, in company with the bishop, to
visit them. They related their story, and shortly after expired.

Thus many chroniclers narrate of the seven sleepers of Ephesus.
All are not agreed as to the place where this extraordinary event occurred. It has been assigned also to the "mountain of the seven sleepers," near Tersous. It may have been claimed by the citizens of twenty
other ancient cities, for aught we can tell: Faith removes mountains.

But the number remains intact. Mahomet wrote of seven heavens—no Mahometan takes the trouble to believe in less. The "wise men were but seven"; there were seven poets of the age of Theocritus; seven of the daughters of Pleione elevated to the back of Taurus; and

> There were seven pillars of gothic mould,
> In Chillon's dungeon, dark and old,

and wherefore not *seven* sleepers at Ephesus or Tersous; or seven sisters of

> Nature's sweet restorer, balmy sleep?

Although not to be found in Livy, or Hesiod, or Ovid, or any of the fathers of history or fable, there is a legend of the latter *seven*, which may be considered in the light of an abstract of title of certain seven sisters, to be included in the list of immortal sevens who have honoured the earth by making it their abode.

It is many thousands of years since Sleep received from her parent, as a dowry of love, an empire, unequalled in extent by any other which the earth ever acknowledged. Her domain embraced "the round world, and they that dwell therein." From pole to pole, and from ocean to ocean, she swayed her sceptre. And it was assigned her that man should devote one-third of his existence in paying homage at the foot of her throne. All monarchs from Ninus to Napoleon have done her honour. All ladies from Rhodope to Cleopatra, and from Helen to Clothilde, have admitted her claim to ascendency. All serfs, and all captives, from Epictetus to Abd-el-Kader, have forgotten their bonds and their captivity, and bowed, on an equality with kings, beneath her nod.

Sleep had seven sisters. Envious of her throne, and jealous of her power, they complained bitterly that no heritage, and no government, and no homage was theirs. Then they strove to deceive men, and counterfeit the blessings which Sleep conferred, and thus to steal the affections of her subjects from the universal monarch, and transfer them to themselves. Herein they toiled and invented many strange devices; and though they beguiled many, these all fell back again to the allegiance they had sworn of old.

"O my sisters!" said Sleep, "wherefore do you strive to instil discontent into the hearts of my subjects and breed discord in my dominions? Know ye not, that all mortals must fain obey me, or die? Your enchantments cannot diminish my votaries, and only serve to increase my power. And men, who for a while are cheated of the blessings I confer, woo me at last with increased ardour, and with songs of gratitude fall at my feet."

Morphina first replied—

"We know full well, proud sister, how wide is your empire, and how great your power, but we too must reign, and our kingdoms will soon compare with yours. Let us but share with you in ruling the world, or we will rule it for ourselves."

"Sisters! let us be at peace with each other. Is there not two-thirds of the life of man free from my control? Why should you not steal from iron-handed care enough of power to make you queens as potent, or little less than me? My minister of dreams shall aid you by his skill, and visions more gorgeous, and illusions more splendid, than ever visited a mortal beneath my sway, shall attend the ecstacies of your subjects."

The sisters were reconciled henceforth. And anon thousands and millions of Tartar tribes and Mongolian hordes welcomed Morphina, and blessed her for her soothing charms and benignant rule—blessed her for her theft from the hours of sorrow and care—blessed her for the marvels of dreams the most extravagant, and visions the most gorgeous that ever arose in the brain of dweller in the glowing East.

More extended became the sway of the golden-haired Virginia, until four-fifths of the race of mortals burned incense upon her altars, or silently proffered thank-offerings from their hearts. Curling ever upwards from the hearth of the Briton and the forest of the Brazilian—from the palaces of Ispahan and the wigwams of the Missouri—from the slopes of the eternal hills and the bosom of the mighty deep, arose the fragrant odours of her votaries, mingled with the hum of pæans in her praise.

Beneath the shadow of palms, in the sultry regions of the sun, the dark impetuous Gunja held her court. There did the sons of the Ganges and the Nile, the Indus and the Niger, own her sovereignty; and there did the swarthy Hindoo and the ebon African hold festivals in her honour. And, though the hardy Norseman scorned her proffered offices, she established her throne in millions of ardent and affectionate hearts.

Not far away, the red-lipped Siraboa raised her graceful standard from the summit of a feathery palm; and the islanders of the Archipelago, in proa and canoe, hastened to do her homage. The murderous Malay stayed his uplifted weapon, to bless her name; and savage races, that ne'er bowed before, fell prostrate at her feet.

Honoured by the Incas, and flattered by priests—persecuted by Spanish conquerors, but victorious, Erythroxylina established herself in the Bolivian Andes and the Cordilleras of Peru. With subjects the most devoted and faithful, she has for ages received the homage of a kingdom of enthusiastic devotees.

Two, less favoured, less beautiful, and less successful of the sisters, pouting and repining at the good fortune that had attended the others, secluded themselves from the rest of the world, and rushed into voluntary exile. Datura, ruddy as Bellona, fled to the Northern Andes; and in those mountainous solitudes collected a devoted few of frantic followers, and established a miniature court. The pale and dwarfish Amanita, turning her back on sunny lands and glowing skies, sought and found a home and a refuge, a kingdom and a court, in the frozen wastes of Siberia.

And now in peace the sisters reign, and the world is divided between them. When care, or woe, or wan disease, steals for a time the mortal from his allegiance to the calm and blue-eyed Sleep, then do the sisters ply their magic arts to win him back again, and, by their soothing influence, lull him to rest once more, and again unlock the portals of the palace of dreams; then issues from the trembling lips the half-heard murmur of a whispered blessing on the *Seven Sisters of Sleep*.*

In all times Sleep has been a fertile theme with poets—one on which the best and worst has been written. All forms in heaven and in earth have submitted themselves to become similes; and columns

---

*The learned in the lore of ancient Rome may charge us, if they will, with a grievous wrong in considering Sleep as one of the softer sex, inasmuch as Somnus was one of the elder of the "*lords* of the creation." We confess to an inclination towards the "*ladies* of creation;" and in this matter especially
  We have a vision of our own
  And why should we undo it?

of adjectives have done duty in the service since Edmund Spencer raised his House of Sleep, where

> *careless Quiet lyes,*
> *Wrapt in eternal silence, farre from enimyes.*

No monarch has numbered so many odes in his praise, or had so many poet laureates "all for love." These, though not so long, are quite as worthy as the one we heard when George III was no longer king. Perhaps that same little tyrant, LOVE, has come in for even a larger share of what some would call "twaddle." In the sunny morn of youth, these hung upon our lips, and dwelt in our hearts, with less of doubt than disturbs their present repose. Old age makes us sleepy, and we sing—

> *O magic sleep! O comfortable bird,*
> *That broodest o'er the troubled sea of the mind*
> *Till it is hushed and smooth! O unconfined*
> *Restraint, imprisoned liberty, great key*
> *To golden palaces, strange minstrelsy,*
> *Fountains grotesque, new trees, bespangled caves,*
> *Echoing grottoes, full of tumbling waves*
> *And moonlight; aye, to all the mazy world*
> *Of silvery enchantments!*
>
> Endymion

"God gave sleep to the bad," said Sadi, "in order that the good might be undisturbed." Yet to good and bad sleep is alike necessary. During the hours of wakefulness the active brain exerts its powers without cessation or rest, and during sleep the expenditure of power is balanced again by repose. The physical energies are exhausted by labour, as by wakefulness are those of the mind; and if sleep comes not to reinvigorate the mental powers, the overtaxed brain gives way, and lapses into melancholy and madness. Men deprived of rest, as a sentence of death, have gone from the world raving maniacs; and violent emotions of the mind, without repose, have so acted upon the body, that, as in the case of Marie Antoinette, Ludovico Sforza, and others, their hair has grown white in a single night—

As men's have grown from sudden fears.*

Mind and body alike suffer from the want of sleep, the spirit is broken, and the fire of the ardent imagination quenched. Who can wonder that when disease or pain has racked and tortured the frame, and prevented a subsidence into a state so natural and necessary to man, he should have resorted to the aid of drugs and potions, whereby to lull his pains, and dispel the care which has banished repose, and woo back again—

the certain knot of peace,
The baiting place of with, the balm of woe;
The poor man's wealth, the prisoner's release,
Th'indifferent judge between the high and low.

Leigh Hunt has well said, "It is a delicious moment that of being well nestled in bed, and feeling that you shall drop gently to sleep. The good is to come, not past; the limbs have just been tired enough to render this remaining in one posture delightful; the labour of the

---

*A correspondent of the Medical Times having asked for authentic instances of the hair becoming grey within the space of one night, Mr. D. F. Parry, Staff-Surgeon at Aldershott, transmitted the following account, of which he made memorandum shortly after its occurrence. "On February 19, 1858, the column under General Franks, in the south of Oude, was engaged with a rebel force at the village of Chamda, and several prisoners were taken. One of them, a sepoy of the Bengal army, was brought before the authorities for examination, and I, being present, had an opportunity of watching from the commencement the fact I am about to record. Divested of his uniform, and stripped completely naked, he was surrounded by the soldiers, and then first apparently became alive to the danger of his position; he trembled violently, intense horror and despair were depicted in his countenance, and although he answered all the questions addressed to him, he seemed almost stupified with fear; while actually under observation, within the space of half-an-hour, his hair became grey on every portion of his head, it having been, when first seen by me, the glossy jet black of the Bengalee, aged about twenty-four. The attention of the bystanders was first attracted by the serjeant, whose prisoner he was, exclaiming, 'He is turning grey'; and I, with several other persons, watched its progress. Gradually, but decidedly, the change went on, and a uniform greyish colour was completed within the period above named."

day is gone—a gentle failure of the perceptions creeps over you—the spirit of consciousness disengages itself once more, and with slow and hushing degrees, like a mother detaching her hand from that of a sleeping child, the mind seems to have a balmy lid closing over it, like the eye—it is closed—the mysterious spirit has gone to take its airy rounds."

It is this universal sense of the blessing of sleep which takes hold of the mind with such a religious feeling, that the appearance of a sleeping form, whether of childhood or age, checks our step, and causes us to breathe softly lest we disturb their repose. We can scarce forbear whispering, while standing before the well-known picture of the "Last Sleep of Argyle," lest by louder or more distinct articulation, we should rob the poor old man of a moment of that absence of sorrow which sleep has brought to him for the last time.

Shakespeare has made the murder of Duncan to seem more revolting in that it was committed while he slept. Macbeth himself must have felt this while exclaiming—

Methought I heard a voice cry, "Sleep no more!
Macbeth does murther sleep, the innocent sleep;
Sleep, that knits up the ravelled sleave of care,
The death of each day's life, sore labour's bath,
Balm of hurt minds, great nature's second course,
Chief nourisher in life's feast."

Had Desdemona been sent to her last account at once, when her lord entered the room and kissed her as she slept, we feel that all our pity for the jealous Moor would have been turned to hate, and our detestation of him been so great that no room had been left for execration of the villanous Iago, who now seems to be the Mephistopheles, the evil genius of the work.

"A blessing," says Sancho Panza, "on him who first invented sleep; it wraps a man all round like a cloak." But neither Sancho nor any one else will give us a blessing if we suffer ourselves to go to sleep in thinking over it, at the very threshold of our enterprise, and before indulging in communion with the seven sisters of whom we have spoken. It was a trite remark of a divine that "where drowsiness begins, devotion ends," and needs application as much to book writers as to sermon preachers. Although we may not have the power to check an occasional yawn, in which there may be as much temporal relief as in

a good sneeze, let us avoid the premonitory sinking of the upper eyelids, by calling in the aid of Francesco Berni to release us from the
spell of sleep, and introduce us to "the sisters" of the olden time.

Quella diceva ch'era la piu bella
Arte, il piu bel mestier che si facesse;
Il letto er'una veste, una gonella
Ad ognun buona che se la mettesse.

        Orland. Innamor, lib. iii. cant. vii

CHAPTER II

THE SISTERS OF OLD

What are these,
So withered, and so wild in their attire;
that look not like the inhabitants o' the earth,
And yet are on't?

Macbeth

here is no reason to doubt that the ancients were, in a manner, acquainted with some of the narcotics known to us, although they did not indulge in them as stimulants or luxuries. The antiquarian, it is true, has failed to unearth the tobacco-box of Claudius, or the pipe of Nero—however much the latter may have been given to smoke. And no one has as yet discovered a snuff-box bearing the initials of Marc Antony, whence the taper fingers of Egypt's queen drew a pinch of Princess' Mixture or Taddy's Violet, gazing with loving eyes on Antony the while. In those remote times the hemp and the poppy were not unknown; and there is reason for believing that in Egypt the former was used as a potion for soothing and dispelling cares.

Herodotus informs us that the Scythians cultivated hemp, and converted it into linen cloth, resembling that made from flax; and he adds also, that "when, therefore, the Scythians have taken some seed of this hemp, they creep under the cloths, and then put the seed on the red hot stones; but this being put on smokes, and produces such a steam, that no Grecian vapour-bath would surpass it. The Scythians, transported with the vapour, shout aloud."* The same author also states

---

*Herod., lib. iv. cap. 74–75.

that the Massagetæ, dwelling on an island of the Araxes, have discovered "trees that produce fruit of a peculiar kind, which the inhabitants,
when they meet together in companies, and have lit a fire, throw on
the fire as they sit round in a circle; and that by inhaling the fumes of
the burning fruit that has been thrown on, they become intoxicated
by the odour, just as the Greeks do by wine, and that the more fruit is
thrown on, the more intoxicated they become, until they rise up to
dance, and betake themselves to singing."*

Homer also makes Helen administer to Telemachus, in the house
of Menelaus, a potion prepared from *nepenthes*, which made him forget his sorrows.

> *Meanwhile with genial joy to warm the soul,*
> *Bright Helen mix'd a mirth-inspiring bowl;*
> *Temper'd with drugs of sovereign use to assuage*
> *The boiling bosom of tumultous rage;*
> *To clear the cloudy front of wrinkled care,*
> *And dry the tearful sluices of despair;*
> *Charm'd with that virtuous draught, the exalted mind*
> *All sense of woe delivers to the wind:*
> *Though on the blazing pile his parent lay,*
> *Or a loved brother groan'd his life away,*
> *Or darling son, oppress'd by ruffian force,*
> *Fell breathless at its feet a mangled corse;*
> *From morn to eve, impassive and serene*
> *The man entranced would view the deathful scene.*
> *These drugs, so friendly to the joys of life,*
> *Bright Helen learn'd from Thone's imperial wife,*
> *Who sway'd the sceptre where prolific Nile*
> *With various simples clothes the fatten'd soil.*
> *With wholesome herbage mixed, the direful bane*
> *Of vegetable venom taints the plain;*
> *From Pæon sprung, their patron-god imparts*
> *To all the Pharian race his healing arts.*
>
> Pope's Homer's Odyssey, b. iv

---

*Ib., lib. i. cap. 202.

Diodorus Siculus states that the Egyptians laid much stress on the circumstance that the plant used by Helen had been given her by a woman of Egyptian Thebes, whence they argued that Homer must have lived amongst them, since the women of Thebes were celebrated for possessing a secret whereby they could dissipate anger or melancholy. This secret is supposed to have been a knowledge of the narcotic properties of hemp. The plant was known to the Romans, and largely used by them in the time of Pliny for the manufacture of cordage, and there is scarce a doubt that they were acquainted with its other properties. Galen refers to the intoxicating power of hemp, for he relates that in his time it was customary to give hemp-seed to the guests at banquets as a promoter of hilarity and enjoyment. Slow poisons and secret poisoning was an art with which the Romans were not at all unfamiliar. What the medium was through which they committed these criminal acts, can only be conjectured from the scanty information remaining. Hemp, or opium, or both, may have had some share in the work, since the poppy was sacred to Somnus, and known to possess narcotic properties.

The latter plant is one of the earliest described. Homer speaks of the poppy growing in gardens, and it was employed by Hippocrates, the father of physic, who even particularizes two kinds, the black and the white, and used the extract of opium so extensively, as to be condemned by his contemporary Diagoras. Dioscorides and Pliny also make mention of it; and from their time, it has been so commonly used, as to be incorporated in all the materia medicas of subsequent medical writers.

Plutarch tells us that a poison was administered to Aratus of Sicyon, not speedy and violent, but of that kind which at first occasions a slow heat in the body, with a slight cough, and then gradually brings on consumption and a weakness of intellect. One time when Aratus spat up blood, he said, "this is the effect of royal friendship." And Quintilian, in his Declamations, speaks of this poison in such a manner as proves that it must then have been well known.

The infamous acts of Locusta are noticed by Tacitus, Suetonius, and Juvenal. This poisoner seems to have been a type of such a character as the traditions of a later age embodied in the person and under the name of Lucretia Borgia.

Agrippina, being desirous of getting rid of Claudius, but not daring to despatch him suddenly, and yet wishing not to leave him time sufficient to make new regulations concerning the succession to the throne, made choice of a poison which should deprive him of his reason and gradually consume him. This she caused to be prepared by an expert poisoner, named Locusta, who had been condemned to death for her infamous actions, but saved that she might be employed as a state engine. The poison was given to the emperor in a dish of mush-rooms, but as, on account of his irregular manner of living, it did not produce the desired effect, it was assisted by some of a stronger nature. We are also further told that this Locusta prepared the drug where-with Nero despatched Britannicus, the son of Messalina, whom his father, Claudius, wished to succeed him on the throne. As this poison occasioned only a dysentery, and was too slow in its operation, the emperor compelled Locusta, by blows, and by threatening her with death, to prepare in his presence one more powerful. It was first tried on a kid, but as the animal did not die till the end of five hours, she boiled it a little longer, until it instantaneously killed a pig to which it had been given, and this poison despatched Britannicus as soon as he had tasted it. For this service the emperor pardoned Locusta, rewarded her liberally, and gave her pupils, whom she was to instruct in her art, in order that it might not be lost.

The pupils of Locusta have not left us, however, the secret which their mistress confided to them. The demand made of the apothecary in *Romeo and Juliet* would have suited Nero's case, in the latter instance.

Let me have
A dram of poison; such soon speeding geer
As will disperse itself through all the veins,
That the life-weary taker may fall dead;
And that the trunk may be discharged of breath
As violently, as hasty powder fired
Doth hurry from the fatal cannon's mouth.

What connection the narcotic hemp had with the famous oracle of Delphi is not altogether certain, but it has been supposed, and such supposition contains nothing of heresy in these days, that the ravings of the Pythia were the consequences of a good dose of haschish, or bang. The non-classical readers will allow us to inform them, and the classical permit us to remind them, that the oracle at Delphi was the most celebrated in all Greece. That it was related of old, that a certain shepherd, tending his flocks on Mount Parnassus, observed, that the steam issuing from a hole in the rock seemed to inspire his goats, and cause them to frisk about in a marvellous manner. That this same shepherd was tempted to peep into the hole himself, and the fumes rising therefrom filled him with such ecstacy, that he gave vent to wild and extravagant expressions, which were regarded as prophetical. This circumstance becoming known, the place was revered, and thereon a temple was afterwards erected to Apollo, and a priestess appointed to deliver the oracles. This priestess of Apollo, Pythia, was seated over the miraculous cavity upon a tripod, or three-legged stool, and the fumes arising were supposed to fill her with inspiration, and she delivered, in bad verses, the oracles of the deity. During the inspiration, her eyes sparkled, her hair stood erect, and a shivering ran over the whole body. Under the convulsions thus produced, with loud howlings and cries, she delivered the messages, which were carefully noted down by an attendant priest. Plutarch states, that one of the priestesses was thrown into such an excessive fury, that not only those who came to consult the oracle, but the priests in attendance, were so terrified, that they forsook her and fled; and that the fit was so violent, that she continued several days in agony, and finally died. It has been believed that these fumes, instead of proceeding from earth, were produced by the burning of some narcotic herb, probably hemp. Who shall decide?

In later times "bang" is referred to in the *Arabian Nights*. In one of the tales, two ladies are in conversation, and one enquires of the other, "If the queen was not much in the wrong not to love so amiable a prince?" to which the other replied, "Certainly, I know not why she goes out every night and leaves him alone. Is it possible that he does not perceive it?" "Alas!" says the first, "how would you have him to perceive it? She mixes every evening with his drink the juice of a certain herb, which makes him sleep so sound all night, that she has time to go where she pleases, and as day begins to appear, she comes to him again, and awakes him by the smell of something she puts under his nose."

The Caliph Haroun al Raschid indulged too in "bang" and although somewhere we have seen this word rendered "henbane," we still adhere to the "bang" of the text, and think the evidence is in favour of the Indian hemp. Further accounts of the early history of this plant we will not however forestall, as it will occur more appropriately when we come to speak of it in particular. Henbane has been long enough known; but is has always had the misfortune either of a positive bad name, or no one would speak much in its favour, and therefore it has never risen in the world.

The lettuce, which has not been known to us three hundred years, was also known to the ancients, and its narcotic properties recognized. Dioscorides writes of it, and so also Theophrastus. It is referred to by Galen, and, if we mistake not, spoken of by Pliny. It was certainly wild, in some of its species, on the hills of Greece, and was cultivated for the tables of the salad-loving Greeks and Romans. It had been better that some of them had spent more of their time eating lettuce salads, and by that means had less time to spare for other occupations of a far more reprehensible kind.

The "nepenthes" of Homer has already been shown to have found a representative in hemp. There have also been claims made for considering it as the crocus, or the stigmas of that flower known to us as saffron. Pliny states that it has the power of allaying the fumes of wine, and preventing drunkenness; and it was taken in drink by great winebibbers, to enable them to drink largely without intoxication. Its properties are of a peculiar character, causing, in large doses, fits of immoderate laughter. The evidence in favour of this being the true "nepenthes" is, however, we consider very incomplete, and not so satisfactory, by any means, as that given on behalf of the Indian hemp.

When the Roman soldiers retreated from the Parthians, under the command of Antony, Plutarch narrates of them that they suffered great distress for want of provisions, and were urged to eat unknown plants. Among others, they met with a herb that was mortal; he that had eaten of it lost his memory and his senses, and employed himself wholly in turning about all the stones he could find, and, after vomiting up bile, fell down dead. Attempts to unravel the mysteries of this plant have ended, in some cases at least, in referring it to the belladonna, a plant common enough in these our days, and known to possess poisonous properties of a narcotico-acrid character.

An analogous circumstance occurred in the retreat of the Ten Thousand, as related by Xenophon. Near Trebizond were a number of beehives, and as many of the soldiers as ate of the honeycombs became senseless, and were seized with vomiting and diarrhœa, and not one of them could stand erect. Those who had swallowed but little looked very like drunken men, those who ate much were like madmen, and some lay as if dying; and thus they lay in such numbers, as on a field of battle after a defeat. And the consternation was great; yet no one was found to have died: all recovered their senses about the same hour on the following day; and on the third or fourth day thereafter, they rose up as if they had suffered from the drinking of poison.

This poisonous property of the honey is said to be derived by the bees from the flowers of a species of rhododendron (*Azalea pontica*), all of which possess narcotic properties.

Supposing that blind old Homer—if ever there was an old Homer, and if blind, no matter—knew the secret of Egyptian Thebes, and the power of the narcotic hemp, and yet never smoked a hubble-bubble, it is of little consequence, except to the Society of Antiquaries, and certainly makes no difference to Homer now. Although Diagoras condemned Hippocrates for giving too much opium to his patients, we are not informed whether it was administered in the shape of "Tinctura opii," or "Confectio opii," or "Extractum opii," or "Godfrey's cordial," or "Paregoric elixir." The discovery would not lengthen our own lives, and therefore we do not repine. We think that we have some consolation left, in that we are wiser than Homer or Hippocrates in respect of that particular vanity, called "shag tobacco," which, we venture to suggest, neither of those venerable sages ever indulged in during the period of their natural lives. And although Herodotus found the Scythians using, in a strange manner, the tops of the hemp plant, he never got so far as Kamtschatka, and therefore never saw a man getting drunk upon a toadstool. If he had ever seen it, he had never slept till he had told it to that posterity which he has left us to enlighten.

CHAPTER III

THE WOND'ROUS WEED

Canst thou not minister to a mind diseased;
Pluck from the memory a rooted sorrow;
Raze out the written troubles of the brain;
And, with some sweet oblivious antidote,
Cleanse the stuff'd bosom of that perilous stuff,
Which weighs upon the heart?

Macbeth

Amongst Mahometans, the following legend is said to be accepted as an account of the miraculous introduction of the "wond'rous weed" to the world.

Mahomet, passing the desert in winter, found a poor viper frozen on the ground; touched with compassion, he placed it in his sleeve, where the warmth and glow of the blessed body restored it to life. No sooner did the ungrateful reptile find its health restored, than it poked forth its head, and said—

"Oh, Prophet, I am going to bite you."

"Give me a sound reason, O snake, and I will be content."

"Your people kill my people constantly, there is war between your race and mine."

"Your people bite my people, the balance between our kindred is even, between you and me; nay, it is in my favour, for I have done you good."

"And that you may not do me harm, I will bite you."

"Do not be so ungrateful.'

"I will! I have sworn by the Most High that I will."

At the Name the Prophet no longer opposed the viper, but

16

bade him bite on, in the name of God. The snake pierced his fangs in the blessed wrist, which the Prophet not liking, shook him off, but did him no further harm, nor would he suffer those near him to destroy it, but putting his lips to the wound, and sucking out the poison, spat it upon the earth. From these drops sprang that wond'rous weed, which has the bitterness of the serpent's quelled by the sweet saliva of the Prophet.*

Happy Moslem! you have solved the mystery, and your heart feels no doubt; but Christian dogs despairingly sigh for some revelation from the past, whether through history or tradition, of the first use of this plant. In vain we enquire who it was that first conceived and put in practice the idea of burning the large leaves of a weed, and drawing in the smoke to spit it out again? Who it was that discovered pleasure or amusement in tickling the nose with that "titillating dust" to enjoy the luxury of a sneeze, or find employment in blowing it out again? Ye shades of heroes departed, that hover around the pine-woods of the Saskatchewan, sail over the rolling prairies of Illinois, or roam along the strands of Virginia, tell us to what illustrious progenitor of Cree or Mohawk we are to accord the honour of a discovery more popular than any since the days when "Adam delved and Eve span?"

In default of the shades giving us the required information, we must resort to the faint footsteps which "the habit" has left imprinted on the sands of Time. Even the name by which it is called, has been disputed and even denied, as of right, belonging to tobacco. This word, Humboldt informs us, like the words *savannah, maize, maguey,* and *manati,* belong to the ancient language of Hayti or St. Domingo, and did not properly denote the herb, but the pipe through which it was smoked. Tobacco, according to Oveido, was indigenous in Hispaniola, and much used by the native Indians, who smoked it from a tube in the shape of the letter Y, the two branches being inserted in the nostrils, and the stem placed in the burning leaves. The plant was called the *cohiba,* and the rude instrument by which it was inhaled *tabaco.*

Other fabulous accounts of the origin of this mystic name, which opens the heart and hand of the savage more readily than that of gold, trace it to Tabacco, a province of Yucatan in New Spain, whence it is

---

The Ansayrii and the Assassins, by the Hon. F. Walpole.

stated to have been first brought to Europe. Or affinity is claimed for it with the Island of Tobago, one of the Caribbees, where it grew wild in abundance. Or its derivation is traced to Tobasco, in the island of Florida. In Mexico it was called *yetl*, and in Peru *sagri*, meaning in those languages "the herb," or the herb *par excellence*, worthy of superiority over all other herbs which the earth ever produced from her bosom.

It seems surprising that a vegetable production so universally spread should have different names among neighbouring people. In North America the Algonkin name is *sema*, and the Huron *oyngoua*, and the same dissimilarity exists in the languages of South-American tribes; the Omagua, *petema*; the Maypure, *jema*; the Chiquito, *pâis*; the Vilela, *tusup*; and the Tamanac, *cavai*. One would have expected to have found names with less variation among such neighbours. It might be urged, perhaps, that these are all independent ancient names given by each tribe to the plant before they became acquainted with the existence of their neighbours, and an evidence that its use was not derived from each other, nor from travellers passing among them. To these speculations the theorist is welcome.

There is little reason to doubt that tobacco is a plant indigenous to the New World. With the era, therefore, of Columbus, our knowledge of it will necessarily commence. When the Spaniards landed with that navigator in Cuba in 1492, they found the Cubans doing the same kind of thing as the voyager would now find them occupied in, making and smoking cigars. In the latter act, these Spaniards soon followed the Cuban example, as did those also who landed in 1518, with Fernando Cortez, in the island of Tobago, to a still greater extent. The honour of introducing this, the fairest of "the Seven Sisters of Sleep," to European society and soil, is due, perhaps, to Hernandez, the naturalist, who brought the first seeds from Mexico (Humboldt states, from the Mexican province of Yucatan), in 1559, and conveyed them to Spain. About the same time some unknown Flamingo introduced the illustrious visitor to Portugal.

Of the introduction of tobacco into France, the more commonly-received opinion is, that the first seeds were sent to Catherine de Medici from Portugal in 1560, by Jean Nicot, the French ambassador to that country, and ever since it has borne as its generic name a memento of its patron. Other accounts attribute to Father Andrè Thevet, or some friend of his, the honour of introducing the raw material to the most accomplished snuff-takers in Europe, and, perhaps, the first who ever indulged in it to any extent.

In Tuscany, tobacco was first cultivated under Cosmo de Medici, who died in 1574. It was originally raised by Bishop Alfonso Tournabuoni, from seeds received from his nephew, Nicolo Tournabuoni, then ambassador at Paris. After him it bore the name of Erba Tournabuoni, as in France it was called Herbe de la Reine. Very early, before 1589, the Cardinal Santa Croce, returning from his nunciature in Spain and Portugal to Italy, carried with him thither tobacco; but he can scarce claim the honour of its introduction, although the exploit was commemorated by Castor Duranti in Latin verse. Thus it would appear that this plant was brought from Mexico to Spain, whence it passed into France, and thence into Italy, during the early part of the latter half of the sixteenth century.

The first introduction of tobacco into England has been claimed for a trinity of valiant knights—Sir Francis Drake, Sir John Hawkins, and Sir Walter Raleigh. In Bancroft's *History of the United States*, it is said—"The exiles of a year had grown familiar with the favourite amusement of the lethargic Indians, and they introduced into England the general use of tobacco." These exiles were brought home by Drake before Raleigh visited the New World, and the period for the introduction of tobacco into this country by Sir Francis, claims the date of 1560. For Sir John Hawkins' introduction, the time has been fixed at 1565; whilst the earliest date assigned for its introduction by Sir Walter Raleigh is 1584, the same year in which a proclamation was issued in England against it. Humboldt states that the celebrated Raleigh contributed most to introduce the custom of smoking among the nations of the North. When Raleigh brought tobacco from Virginia to England, whole fields of it were already cultivated in Portugal. It was also previously known in France, where it was brought into fashion by Catherine de Medici. As early as the end of the sixteenth century, bitter complaints were made in England of this imitation of the manners of a savage people. It was feared, that by the practice of smoking tobacco, Englishmen would degenerate into a barbarous state.* The cultivation of

---

* Ex illo sane tempore [tabacum] usu cepit esse creberrime in Angliâ et magno pretio dum quam plurimi graveolentem illius fumum per

this narcotic plant preceded that of the potato in Europe 120 or 140 years.

Camden, who informs us of these fears for the civilization of England, also states that Richard Fletcher, Bishop of London, a courtly prelate (who died in 1596), by the use of tobacco "smothered the cares he took by means of his unlucky marriage." According to Aubrey, the pipe was handed from man to man round the table; and this bears, certainly, a great resemblance to the custom of the North-American Indians—the chief smoking two or three whiffs, then passing it to his neighbour, until from one to another it passes round the circle, and comes back to the first smoker again.

M. Jorevin, a Frenchman, who visited England in Charles II's time, says that the women smoked tobacco as well as the men.

From England the practice of smoking was carried to the Continent. Dutch students were first taught the art of smoking at the University of Leyden by students from England; hence the greatest smokers in Europe derived their knowledge of the use of the pipe from the English.

Lilly, in his autobiography, informs us that when committed to the guard-room in Whitehall, he thought himself in regions far below, where Orpheus sang, and Pluto reigned, for "some were sleeping, others swearing, others smoking tobacco; and in the chimney of the room were two bushels of broken tobacco-pipes." Good friend Lilly, what wouldst thou have thought of a visit to a Studenten Kneipe, where a crowd of students, amid fumes dense as a London fog in November, scream and growl the well-known song—

> And smokes the Fox tobacco?
> And smokes the Fox tobacco?
> And smokes the leathery Fox tobacco?
> Sa! Sa!
> Fox tobacco.
> And smokes the Fox tobacco.
> Then let him fill a pipe!
> Then let him fill a pipe!

---

tubulum testaceum hauriunt et mox e naribus effiant; adeo ut Anglorum corporum in barbarorum naturam degenerasse videantur quum iidem ac barbari delectentur.

—Camden, *Annal. Elizab.*, p. 143 (1585)

Then let him fill a leathery pipe;
Sa! Sa!
Leathery pipe.
Then let him fill a pipe!

And then perhaps—but let the reader enquire for himself of some descendant from the ancestors of the renowned Wouter Van Twiller, the worthy head of the long-pipe faction. In 1601, tobacco was carried to Java, whence it spread over the East. It was also conveyed to Turkey and Arabia in the beginning of this century. El-Is-hakee states that the custom of smoking tobacco began to be common in Egypt between the years of the flight, 1010 and 1012 (A.D. 1601–1603). And from Persian writers on materia medica, it appears to have been introduced into India in A.H. 1014 (A.D. 1605), towards the end of the reign of Jelaladeen Akbar Padshaw. From India, tobacco probably found its way to the Malayan Peninsula and China; although Pallas, Loureiro, and Rumphius think that tobacco was known in China before the discovery of the New World, and that the Chinese tobacco plant is indigenous to that country.

From *Notes and Queries* we learn that "tobacco was first cultivated in this country at Winchcombe in Gloucestershire, and that the natives did suck thereout no small advantage; and before the time of James II the best Virginia was but two shillings the pound, and two gross of the best glazed pipes, and a box with them, three shillings and fourpence." Tobacco became almost a necessary among the upper classes; nor could the parliamentary representatives of the city of Worcester be despatched up to town until the "collective wisdom" had smoked and drunk sack at the "Globe," or some other hostelry. As early as 1621, it was moved in the House of Commons by Sir William Stroud, that "he would have tobacco banished wholly out of the kingdom, and that it may not be brought from any part, nor used amongst us." And by Sir Grey Palmes, "that if tobacco be not banished, it will overthrow 100,000 men in England, for it is now so common, that he hath seen ploughmen take it as they are at the plough." At a later period of the same century, so inveterate had the practice become, that an order appears on the journals of the House, "That no member in the House do presume to smoke tobacco in the gallery, or at the table of the House sitting at Committees."

But tobacco did not come into general use in Europe without great and strenuous opposition. All kinds of weapons were called in requisition to stay its progress. Persuasion and force were alike essayed without effect. A German writer has collected the titles of a hundred different works condemning its use, which were published within half a century of its introduction into Europe. The pen was wielded by royal as well as plebeian fingers, and the famous diatribe of the British Solomon, King James I, of blessed memory, defender of the faith, and antagonist of tobacco, keeps his memory still *green* in the hearts of Englishmen. In Russia, the snuff-taker was ingeniously cured of the habit, by having his nose cut off, while smokers had a pipe bored through the same useful projection. Michael Feodorovitch Tourieff kindly offered a bastinado to the Muscovites for the first offence, cutting off the nose for the second, and the head for the third. In 1590, Pope Innocent XII took the trouble to excommunicate all who used tobacco in any form in the church of St. Peter's in Rome. And in 1624, Pope Urban VII, the old woman, fulminated a bull against all persons found taking snuff during divine service; and old women, in the spirit of opposition, have been fond of snuff ever since. The sultans and priests of Persia and Turkey declared smoking a sin against their religion. Amurath IV of Persia published an edict, making the smoking of tobacco a capital offence. Shah Abbas II punished such delinquents equally severely. When leading an army against the Cham of Tartary, he proclaimed that every soldier in whose possession tobacco was found, would have his nose and lips cut off, and afterwards be burnt alive. El-Gabartee relates, that about a century ago, in the time of Mohammed Básha El-Yedek-shee, who governed Egypt in the years of the flight, 1156–58, it frequently happened that, when a man was found with a pipe in his hand in Cairo, he was made to eat the bowl with its burning contents. This may seem incredible, but a pipe bowl *may* be broken by strong teeth, particularly if it be of meerschaum. In Tuscany, the growth of tobacco was prohibited, except in a few localities, where it was allowed, under certain restrictions, from 1645 to 1789, when the Grand Duke Peter Leopold declared its cultivation free all over the country. Ferdinand III afterwards restricted it to its former localities. The number of these were reduced in 1826, and in 1830 its growth was entirely prohibited. In Transylvania the penalty for growing tobacco was a total confiscation of property; and for the use of the weed, a fine of from three to two hundred florins. In 1661, the Canton of Berne introduced an eleventh commandment to the decalogue, and this was inserted after the seventh, "Thou shalt not smoke!" In 1719, the wise senate of Strasburg prohibited the cultivation of tobacco, fearing lest it should interfere with the growth of corn. Prussia and Denmark contented themselves with prohibiting its use. This brings us back again to England, and the days of "good Queen Bess." That lady, who is said to have prohibited the use of tobacco in churches, according to certain chroniclers, was wont to banter Sir Walter Raleigh on his affection for his *protégé*. It is said, that on one occasion, when Raleigh was conversing with his royal mistress upon the singular properties of this new and extraordinary herb, he assured her Majesty that he had so well experienced the nature of it, that he could tell her of what weight even the smoke would be in any quantity proposed to be consumed. Her Majesty, deeming it impossible to hold the smoke in a balance, must needs lay a wager to solve the doubt. Raleigh procured the quantity agreed upon, he thoroughly smoked it, and weighed the ashes, pleading at the same time that the weight now wanting was the weight of the smoke dissipated in the process. The Queen did not deny the doctrine of her favourite saying "that she had often heard of those who had turned their gold into smoke, but Raleigh was the first who had turned his smoke into gold."

The Star Chamber levied a heavy duty, and Charles II prohibited its cultivation in England. Tobacco was first put under the excise in 1789. It was not at first allowed to be smoked in ale-houses. "There is a curious collection of proclamations, etc.," says Brand, "in the archives of the Society of Antiquaries of London. In vol. viii is an ale-house licence, granted by six Kentish justices of the peace, at the bottom of which is the following item, among other directions to the inn-holder:—'Item.—You shall not utter, nor willingly suffer to be uttered, drunke, or taken, any tobacco within your house, cellar, or other place thereunto belonging."

Notwithstanding oppositions, imposts, anathemas, counterblasts, and persecutions, tobacco gradually and rapidly arose in popular esteem. The first house in which it was publicly smoked in Britain was the Pied Bull, at Islington; but this was "alone in its glory" for a very brief period of time. "Is it not a great vanity," saith Royal James, "that a man cannot heartily welcome his friend now, but straight they must be in hand with tobacco? And he that will refuse to take a pipe of tobacco amongst his fellows is accounted peevish, and no good company; yea, the mistress cannot in a more mannerly kind entertain her servant than by giving him out of her fair hand a pipe of tobacco." Raleigh smoked in his dungeon in the Tower, while the headsman was grinding his axe. Cromwell loved his pipe, and dictated his despatches to Milton over some burning Trinidado, or sweet-smelling nicotine. Ben Johnson affirmed that tobacco was the most precious weed that the earth ever tendered to the use of man. Dr. Radcliffe recommended snuff to his brethren. Dr. Johnson kept his snuff in his waistcoat pocket; and so did Frederick the Great. Robert Hall smoked in his vestry, and, it would seem, in other places, as well, for Gilfillan informs us, that when on a visit to a brother clergyman, he went into the kitchen where a pious servant girl, whom he loved, was working. He lighted his pipe, sat down, and asked her—"Betty, do you love the Lord Jesus Christ?" "I hope I do, sir," was the reply. He immediately added, "Betty, do you love me?" They were married. And Napoleon took rappee by the handful. And poets wrote, and minstrels sang, in the praise of the "Divine Virginia."

> Thou glorious weed of a glorious land,
> I would not be freed from thy magical wand—
> Though a slave to thy fetters, and bound in thy chain,
> Despairing of freedom, I cannot complain.
>
> Tobacco, I love thee—I bow at thy shrine!
> The longer I prove thee, the less I repine.
> The affection I cherish, no time can assuage—
> Thy joys do not perish, like others, with age.

The mailed Spaniard and red-plumed Indian have fought around it; and gold-seekers have drenched it with the gore of negroes. One whole continent has been enriched by it; and to cultivate it, another continent has been depopulated. Negroes have prayed to their fetishes beside it—many a Cacique now dead smoked it at the war-council, and many a grave, grey-bearded Spaniard, who had fought at Lepanto, or bled in the Low Countries. Old soldiers of Cromwell have smoked it; and while Indians have bartered their gold for English beads, the swarthy Buccaneers looked on, handling their loaded muskets. Tobacco was for some time used as currency in Virginia, as, according to Mr. Galton, is the case now among the Damaras, Ovampo, and other tribes of South-Western Africa.

Forty varieties of tobacco have been described; but the differences are mainly the result of climate, and the mode of culture. It grows well in almost every part of the world. The northern limit in Scandinavia is 62°–63° N. L. The different parts of America in which it is grown include Canada, New Brunswick, United States, Mexico, the Western Coast, as far as 40° S. L. In Africa it is cultivated by the Red Sea and Mediterranean, in Egypt, Algeria, the Canaries, the Western Coast, the Cape, and numerous places in the interior. In Europe, it has been raised successfully in almost every country; in Hungary, Germany, Flanders, and France, it forms an important agricultural product. In Asia, it has spread over Turkey, Persia, India, Thibet, China, Japan, the Philippines, Java, and Ceylon. In parts of Australia and New Zealand. From the Equator to 50° N. L., it may be raised without difficulty. The finest qualities are raised between 15° and 35° N. L.

The most noted tobacco is that of Cuba; and the most extensive growers are the Americans of the United States. Two-thirds of our supply is doubtless derived from the latter source.

In 1665, Virginia exported to England 60,000 pounds. Twenty-five years afterwards, our total imports were double that amount; while in 1858, they amounted to 62,217,705 pounds, including snuff and cigars; hence, we may fairly calculate that, in Great Britain, eight millions of pounds sterling are annually spent in tobacco.

It has been computed that eight hundred millions of the human race are consumers of tobacco, and that the average annual consumption is 70 ounces per head. The total consumption would, therefore, approximate to two millions of tons. The average annual consumption of every male over eighteen years of age, in each of the following countries of Europe, as collected from returns, is, in Austria, 108 ounces; Zollverein, 156 ounces; Steurverein, including Hanover and Oldenburg, 200 ounces; France, 88 ounces; Russia, 40 ounces; Portugal, 56 ounces; Spain, 76 ounces; Sardinia, 44 ounces; Tuscany, 40 ounces; the Papal States, 32 ounces; England, 66 ounces; Holland, 132 ounces; Belgium, 144 ounces; Denmark, 128 ounces; Sweden, 70 ounces; and Norway, 99 ounces. In the United States of America, the consumption is 122 ounces; and in New South Wales, where there are no restrictive duties, it is declared to exceed 400 ounces.

Jamie, thou shouldst been living at this hour,
   Europe hath need of thee.

To what a height of royal indignation the "Misocapnos" would have risen, had its author postponed its publication 250 years, and reappeared, a "new avater," to see it through the press in these latter days. He had then required no Spanish matches to set him on fire; and the "horrible Stygian smoake" would have required the addition of all Catesby's gunpowder to have made the simile worthy of its royal master, unless, peradventure, the weight of five millions of golden sovereigns from the Inland Revenue Office had pressed heavily upon his conscience, and he had purchased himself a new pair of silk stockings, and rested in peace; then he could have returned the old pair he borrowed in his Scotch capital, in which to meet his English Court at London.

Since the days when the green leaf of tobacco was used as a sovereign application for wounds and bruises and the bites of poisonous serpents, there has been no more singular use discovered for any part of this plant than that of certain African tribes, who, Denham says, "colour their teeth and lips with the flowers of the goorjee tree and the tobacco plant. The former, he saw only once or twice; the latter, was carried every day to market at Bornou, beautifully arranged in large baskets. The flowers of both these plants rubbed on the lips and teeth give them a blood-red appearance, which is there thought a great beauty." That the poison of tobacco should have been turned to account is not surprising; and we are more prepared to hear of the bushmen of South Africa poisoning the heads of their arrow, not with nicotine, but with a poison taken from the head of the yellow serpent. These serpents they kill with the oil of tobacco, one drop or two producing spasms and death. Count Bocarmè effectually settled the question of the poisonous property of nicotine, some years since at Mons. It remained for future experimentalists to discover that as well as a *bane*, tobacco was an *antidote*.

A young lady in New Hampshire fell into the mistake of eating a portion of arsenic, which had been prepared for the destruction of rats. Painful symptoms soon led to the discovery. An elderly lady, then present, advised that she should be made to vomit as speedily as possible, and as the unfortunate victim had always exhibited a loathing for tobacco in any shape, that was suggested a ready means of obtaining the desired end. A pipe was used, but this produced no nausea. A large portion of strong tobacco was then chewed, and the juice swallowed, but even this produced no sensation of disgust. A strong decoction was then made with hot water, of this she drank half a pint without producing nausea or giddiness, or any emetic or cathartic action. The pains gradually subsided, and she began to feel well. On the arrival of physicians, an emetic was administered. The patient recovered, and no ill consequences were experienced. Another case occurred a few years subsequent at the same place, when tobacco was administered and no other remedy. In this instance there was complete and perfect recovery. From this it may be reasonably concluded, that tobacco is an antidote of very safe and ready application in cases of poisoning by arsenic.

Financiers and Chancellors of Exchequers or Ministers of Finance, look with particularly favourable eyes upon the "Indian Weed." Our own official in that department, can now calculate on nearly six millions of safe income in his estimates for a year, from this fertile source. Our near neighbours of France consider four millions too good an addition to the revenue, to denounce its use. Austria and Spain each manages to supply the state coffers with a million and a half of money from the tobacco monopoly. Russia, the Zollverein, Portugal, Sardinia, and the Papal States, individually realizes from three to four hundred thousands of pounds every year, from the use or abuse of this most popular plant in the world.

Although this habit, in its increase, may cause throbs of ecstatic joy in the breasts of certain officials, there are other sections of society holding antagonistic opinions. The Maine Conference of the Methodist Episcopal Church at a late session, passed the following preamble and resolutions:

> Whereas— The use of tobacco prevails to a prodigious extent in our country, as indicated in the reports of our national treasury, and other authentic documents, from which it appears that over 100,000,000 pounds of this article are consumed in the United States annually, at a cost to the consumers of over 20,000,000 dollars, and whereas, we have reason to believe that its use is rapidly increasing, and that even ministers of the Gospel are becoming,

to a great extent, guilty of this debasing indulgence; therefore—

I. Resolved. That we view these facts as a matter of profound alarm, and such an evil as to demand the serious attention of the Church.

II. Resolved. That we regard the use of tobacco as an expensive and needless indulgence, unfavourable to cleanliness and good manners, unbecoming in Christians, and especially in Christian Ministers, and like the use of alcohol, a violation of the laws of physical, intellectual, and moral life.

III. Resolved. That we will discountenance the use of that injurious narcotic, except as a medicine prescribed by a physician, by precept and example, and by all proper means.

De Lagny states that the "Old Believers" a sect of dissenters from the Greek Church in Russia, look with horror on the use of tobacco. The Wahhabees, a Pharasaical sect of strict Moslems, are rigid in their condemnation of tobacco, and in their adherence to the precepts of the Koran, and the traditions of the Prophet.

There are to be met with nearer home, those who are inveterate against its use, and who willingly join with Cowper in denouncing the

Pernicious weed which banishes for hours,
That sex whose presence civilizes ours.

An occasional pamphlet or letter, makes its way into the hands of speculative publishers or into class papers, giving gratuitous advice, and much denunciatory language, against a habit which is by far too general, and has been tested by too many experiments not to be well known, and equally well understood. These "counterblasts" differ but little from the model one which each would seem to aim at imitating—the quaint expressions, the only redeeming quality in the original, alone being wanting.

"Surely," saith the high and mightie Prince James,

smoke becomes a kitchen farre better than a dining chamber; and yet it makes a kitchen oftentimes in the inward parts of men, soyling and infecting them with an unctuous and oyly kind of soote, as hath been found in some great tobacco takers, that after their death were opened. Now, my good countrymen, let us (I pray you), consider what honour or policie can move us to imitate the barbarous and beastlie manners of the wild, godlesse, and slavish Indians, especially in so vile and filthy a custome. Shall we, that disdain to imitate the manner of our neighbour, France (having the style of the greate Christian kingdome), and that cannot endure the spirit of the Spaniards, (their king being now comparable in largenesse of dominions to the greatest Emperor of Turkey), shall we, I say, that have been so long civill and wealthy in peace, famous and invincible in war, fortunate in both—we that have been ever able to aid any of our neighbours (but never deafened any of their ears with any of our supplications for assistance), shall we, I say, without blushing, abase ourselves so far as to imitate these beastlie Indians, slaves to the Spaniards, the refuse of the worlde, and, as yet, aliens from the holy covenant of God? Why do we not as well imitate them in walking naked as they do, in preferring glasses, feathers, and toys, to gold and precious stones, as they do? Yea, why do we not deny God, and adore the devils, as they do? Have you not, then, reasons to forbear this filthie noveltie, so basely grounded, so foolishly received, and so grosslie mistaken in the right use thereof? In your abuse thereof, sinning against God, harming yourselves both in person and goods, and raking also, thereby, the marks and notes of vanitie upon you, by the custom thereof, making yourselves to be wondered at by all forreine civill nations, and by all strangers that come among you, to be scorned and contemned; a custom loathsome to the eye, hateful to the nose, harmfull to the braine, dangerous to the lungs, and, in the blacke stinking fume thereof, nearest resembling the horrible Stygian smoake of the pit that is bottomless.

Wise and worthy king, adieu. Gold stick, lead the way. We hasten from your royal presence to join the Cabinet of Cloudland. *Viva la Virginie!*

CHAPTER IV

THE CABINET OF CLOUDLAND

A magnificent array of clouds;
And as the breeze plays on them, they assume
The forms of mountains, castled cliffs, and hills,
And shadowy glens, and groves, and beetling rocks;
And some, that seem far off, are voyaging
Their sunbright path in folds of silver.

"Right," said I to myself, as I lay down the volume of Hyperion, in which I had been glancing for repose.

I, too, have a friend, not yet a sexagenary bachelor, but a bachelor notwithstanding. He has one of those well oiled dispositions which turn upon the hinges of the world without creaking, except during east winds, and when there is no butter in the house. The hey-day of life is over with him; but his old age (begging his pardon) is sunny and chirping, and a merry heart still nestles in his tottering frame, like a swallow that builds in a tumble-down chimney. He is a professed Squire of Dames. The rustle of a silk gown is music to his ears, and his imagination is continually lantern-led by some will-with-the-wisp in the shape of a lady's stomacher. In his devotion to the fair sex—the muslin, as he calls it—he is the gentle flower of chivalry. It is amusing to see how quickly he strikes into the scent of a lady's handkerchief. When once fairly in pursuit, there is no such thing as throwing him out. His heart looks out at his eye; and his inward delight tingles down to the tail of his coat. He loves to bask in the sunshine of a smile; when he can breathe the sweet atmosphere of kid gloves and cambric handkerchiefs, his soul is in its element; and his supreme delight is to pass the morning, to use his own quaint language, "in making dodging calls, and wriggling round among the ladies."

Yet there are a few little points in the picture which want retouching, and beyond all, one great omission to be remedied. It is the *pipe*. What would the worthy Abbot be without his pipe? Just as uncomfortable as we should presume a dog to be without his tail. As incomplete as a sketch of Napoleon without his boots and cocked-hat. See him in a cloud, and he seems the very Premier of Cloudland. It was said of Staines, Lord Mayor of London, that he could not forego his pipe long enough to be sworn into office, without a whiff; and a print was published representing his lordship smoking in his state carriage; the sword bearer smoking, the mace bearer smoking, the coachmen smoking, the footmen smoking, the postilions smoking, and—to crown the whole—all the six horses smoking also. The ninth of November on which this event occurred, must needs have been a cloudy day.

Another cloudy day arose upon London when the great plague broke out, and on this occasion, the smoke of tobacco mingled with the gloom. In *Reliquiæ Hearnianæ*, it is stated that "none who kept tobacconist's shops had the plague. It is certain that smoking was looked upon as a most excellent preservative, insomuch, that even children were obliged to smoke. And I remember" continues the writer, "that I heard formerly Tom Rogers, who was yeoman beadle, say, that when he was that year when the plague raged, a schoolboy at Eton, all the boys of that school were obliged to smoke in the school every morning, and that he was never whipped so much in his life as he was one morning for not smoking." We may imagine the experiences of some of these urchins at their first or second attempt, and in remembrance, it may be, of some similar experience of our own, see no cause for wonder at Tom Rogers not liking to elevate his yard of clay, and view the curls of smoke arise from the ashes of the smouldering weed. Another amateur who flourished after the great fire had burnt out all traces of the great plague, has left us the record of his "day of smoke," and the cudgelling he received for doing that which Tom Rogers was whipped for not doing:

> I shall never forget the day when I first smoked. It was a day of exultation and humiliation. It was a Sunday. My uncle was a great

smoker. He dined with us that day; and after the meal, he pulled out his cigar case, took a cheroot, and smoked it. I always liked the fumes of tobacco, so I went near him and observed how he put the cheroot into his mouth, the way he inhaled the smoke, how he puffed it out again, and the other coquetries of a regular smoker. I envied my uncle, and was determined that I would smoke myself. Uncle fell asleep. Now, thought I, here's an opportunity not to be lost. I quietly abstracted three cigars from the case which was lying on the table, and sneaked off. Being a lad of a generous disposition, I wished that my brothers and cousins should also partake of the benefits of a smoke, so I imparted the secret to them, at which they were highly pleased. When and where to smoke was the next consideration. It was arranged that when the old people had gone to church in the evening, we should smoke in the coach-house. We were six in number. I divided the three cigars into halves, and gave each a piece. Oh, how our hearts did palpitate with joy! Fire was stealthily brought from the cook-house, and we commenced to light our cigars. Such puffing I never did see. After each puff we would open our mouths quite wide, to let the smoke out. At the performance of the first puff we laughed heartily—the smoke coming out of our mouths was so funny. At the second puff we didn't laugh as much, but began to spit; we thought the cigars were very bitter. After the third puff we looked steadfastly at each other—each thought the other looked pale. I could not give the word of command for another pull. I felt choked, and my teeth began to chatter. There was a dead silence for a second. We were ashamed, or could not divulge the state of our feelings. Charlie was the first who gave symptoms of rebellion in his stomach. Then there was a general revolt. What occurred afterwards I did not know, till I got up from my bed next morning, to experience the delights of a sound flagellation. After that I abhorred the smell of tobacco—would never look at a cigar or think of it.

All this happened, as the narrator informed us, at the age of seven—an early age, some may imagine, who do not know that in Vizagapatam and other places on the same coasts, where the women smoke a great deal, it is a common thing for the mothers to appease their squalling brats by transferring the cigar from their own mouths to that of their infants. These youngsters being accustomed to the art of pulling, suck away gloriously for a second, and then fall asleep.

Howard Malcom states,

> that in Burmah the consumption of tobacco for smoking is very great, not in pipes, but in cigars or cheroots, with wrappers made of the leaves of the Then-net tree. In making them, a little of the dried root, chopped fine, is added, and sometimes a small portion of sugar. These are sold at a rupee per thousand. Smoking is more prevalent than "chewing coon" among both sexes, and is commenced by children almost as soon as they are weaned. I have seen,

he continues,

> little creatures of two or three years, stark naked, tottering about with a lighted cigar in their mouth. It is not uncommon for them to become smokers even before they are weaned—the mother often taking the cheroot from her mouth and putting it into that of the infant.

In China, the practice is so universal, that every female, from the age of eight or nine years, as an appendage to her dress, wears a small silken pocket to hold tobacco and a pipe.

The use of tobacco has become universal through the Chinese empire; men, women, children, everybody smokes almost without ceasing. They go about their daily business, cultivate the fields, ride on horseback, and write constantly with the pipe in their mouths. During their meals, if they stop for a moment, it is to smoke a pipe; and if they wake in the night, they are sure to amuse themselves in the same way. It may easily be supposed, therefore, that in a country containing, according to M. Huc, 300,000,000 of smokers, without counting the tribes of Tartary and Thibet, who lay in their stocks in the Chinese markets, the culture of tobacco has become very important. The cultivation is entirely free, every one being at liberty to plant it in his garden, or in the open fields, in whatever quantity he chooses, and afterwards to sell it, wholesale or retail, just as he likes, without the government interfering with him in the slightest degree. The most celebrated tobacco is that obtained in Leao-tong in Mantchuria, and in the province of Sse-tchouen. The leaves, before becoming articles of commerce, undergo various preparatory processes, according to the practice of the locality. In the South, they cut them into extremely fine filaments; the people of the North content themselves with drying them and rubbing them up coarsely, and then stuff them at once into their pipes.

According to etiquette and the custom of the court, Persian princes must have seven hours for sleep. When they get up, they begin to smoke the narghilè or shishe, and they continue smoking all day long. When there is company, the narghilè is first presented to the chief of the assembly, who, after two or three whiffs, hands it to the next, and so on it goes descending; but in general, the great smoke only with the great, or with strangers of distinction. The Schah smokes by himself, or only with one of his brothers, the tombak, the smoke of which is of a very superior kind, the odour being exquisite. It is the finest tombak of Shiraz.

Mr. Neale says, "Talk about the Turks being great smokers; why, the Siamese beat them to nothing. I have often seen a child only just able to toddle about, and certainly not more than two years of age, quit its mother's breast to go and get a whiff from papa's cigaret, or, as they are here termed, borees—cigarets made of the dried leaf of the plaintain tree, inside of which the tobacco is rolled up."

In Japan, after tea drinking, the apparatus for smoking is brought in, consisting of a board of wood or brass, though not always of the same structure, upon which are placed a small fire-pan with coals, a pot to spit in, a small box filled with tobacco cut small, and some long pipes with small brass heads, as also another japanned board or dish, with socano—that is, something to eat, such as figs, nuts, cakes, and sweetmeats. "There are no other spitting pots," says Kœmpfer, "brought into the room but those which come along with the tobacco. If there be occasion for more, they make use of small pieces of bamboo, a hand broad and high being sawed from between the joints and hollowed."

In Nicaragua, the dress of the urchins, from twelve or fourteen downwards, consists generally of a straw hat and a cigar—the latter sometimes unlighted and stuck behind the ear, but oftener lighted and stuck in the mouth—a costume sufficiently airy and picturesque, and excessively cheap. The women have their hair braided in two long locks, which hang down behind, and give them a school-girly look, quite out of keeping with the cool deliberate manner in which they puff their cigars, occasionally forcing the smoke in jets from their nostrils.*

On the Amazon, all persons—men and women—use tobacco in smoking; when pipes are wanting, they make cigarillos of the fine tobacco, wrapped in a paper-like bark, called Toware; and one of these is passed round, each person, even to the little boys, taking two or three puffs in his turn.†

The Papuans pierce their ears and insert in the orifice, ornaments or cigars of tobacco, rolled in pandan leaf, of which they are great consumers.

A Spaniard knows no crime so black that it should be visited by the deprivation of tobacco. In the Havana, the convict who is deprived of the ordinary comforts, or even of the necessaries of life, may enjoy his cigar, if he can beg or borrow it; if he stole it, the offence would be considered venial. At the doorway of most of the shops hang little sheet-iron boxes filled with lighted coals, at which the passer-by may light cigars; and on the balustrade of the staircase of every house stands a small chafing dish for the same purpose. Fire for his cigar, is the only thing for which a Spaniard does not think it necessary to ask and thank with ceremonious courtesy. If he has permitted his cigar to go out, he steps up to the first man he meets—nobleman or galley slave, as the case may be—and the latter silently hands his smoking weed; for it is impossible that two Spaniards should meet and not have one lighted cigar between them. The light obtained, the lightee returns the cigar to the lighter in silence. A short and suddenly checked motion of the hand, as the cigar is extended, is the only acknowledgment of the courtesy. This is never, however, omitted. Women smoke as well as men; and in a full railroad car, every person, man, woman, and child, may be seen smoking. To placard "no smoking allowed," and enforce it, would ruin the road.

A regular smoker in Cuba will consume perhaps twenty or thirty cigars a day, but they are all fresh. What we call a fine old cigar, a Cuban would not smoke.

At Manilla, the women smoke as well as the men. One manufactory employs about 9,000 women in making the Manilla cheroots;

---

*Squier's *Nicaragua*.
† Edwards' *Voyage up the Amazon*.

another establishment employs 3,000 men in making paper cigars or cigarettes. The paper cigars are chiefly smoked by men; the women prefer the "puros," the largest they can get.

The Binua of Johore, of both sexes, indulge freely in tobacco. It is their favourite luxury. The women are often seen seated together weaving mats, and each with a cigar in her mouth. When speaking, it is transferred to the perforation in the ear. When met paddling their canoes, the cigar is seldom wanting. The Mintira women are also much addicted to tobacco, but they do not smoke it.

In South America, many of the tribes are free indulgers in tobacco; and this extends also to the female and juvenile sections of the community. A story, which Signor Calistro narrated to Mr. Wallace whilst travelling in the interior of Brazil, shows that it was nothing but a common occurrence for little girls to smoke. This story is in itself interesting considered apart from all circumstances of veracity.

There was a negro who had a pretty wife, to whom another negro was rather attentive when he had an opportunity. One day the husband went out to hunt, and the other party thought it a good opportunity to pay a visit to the lady. The husband, however, returned rather unexpectedly, and the visitor climbed up on the rafters to be out of sight, among the old boards and baskets that were stowed away there. The husband put his gun by in a corner, and called to his wife to get his supper, and then sat down in his hammock. Casting his eyes up to the rafters, he saw a leg protruding from among the baskets, and thinking it something supernatural, crossed himself, and said, "Lord deliver us from the legs appearing overhead!" The other, hearing this, attempted to draw up his legs out of sight; but, losing his balance, came down suddenly on the floor in front of the astonished husband, who, half-frightened, asked, "Where do you come from?" "I have just come from heaven," said the other, "and have brought you news of your little daughter Maria." "Oh, wife, wife! come and see a man who has brought us news of our little daughter Maria!" then, turning to the visitor, continued, "and what was my little daughter doing when you left?" "Oh, she was sitting at the feet of the Virgin with a golden crown on her head, and smoking a golden pipe a yard long." "And did she send any message to us?" "Oh, yes; she sent many remembrances, and begged you to send her two pounds of your tobacco from the little rhoosa; they have not got any half so good up there." "Oh, wife, wife, bring two pounds of our tobacco from the little rhoosa, for our daughter Maria is in heaven, and she says they have not any half so good up there." So the tobacco was brought, and the visitor was departing, when he was asked, "Are there many white men up there?" "Very few," he replied; "they are all down below with the *diabo.*" "I thought so," the other replied, apparently quite satisfied; "good night."

On the Orinoco, tobacco has been cultivated by the native tribes from time immemorial. The Tamanacs and the Maypures of Guiana wrap maize leaves around their cigars as did the Mexicans at the time of the arrival of Cortes; and, as in Chili, is done at the present day. The Spaniards have substituted paper for the maize husks, in imitation of them. The little cigarettos of Chili are called *hojitas.* They are about two inches and a half long, filled with coarsely powdered tobacco. As their use is apt to stain the fingers of the smoker, the fashionable young gentlemen carry a pair of delicate gold tweezers for holding them. The cigar is so small that it requires not more than three or four minutes to smoke one. They serve to fill up the intervals in a conversation. At tertulias, the gentlemen sometimes retire to a balcony to smoke one or two cigars after a dance.

The poor Indians of the forests of the Orinoco know, as well as did the great nobles of the Court of Montezuma, that the smoke of tobacco is an excellent narcotic; and they use it, not only to procure an afternoon nap, but also to induce a state of quiescence which they call dreaming with the eyes open. At the Court of Montezuma the pipe was held in one hand, while the nostrils were stopped with the other, in order that the smoke might be more easily swallowed. Bernal Diaz also informs us, that after Montezuma had dined, they presented to him three little canes, highly ornamented, containing liquid amber, mixed with a herb they call tobacco, and when he had sufficiently viewed and heard the singers, he took a little of the smoke of one of these canes, and then laid himself down to sleep. A tribe of Indians originally inhabiting Panama, improved upon this method, which occupied both hands, and involved considerable trouble; the method adopted by the chiefs and great men of this tribe, was to employ servants to blow tobacco smoke in their faces, which was convenient

and encouraged their indolence; they indulged in the luxury of tobacco in no other way.

Amongst the Rocky Mountain Indians, it is a universal practice to
indulge in smoking, and when they do so they saturate their bodies in
smoke. They use but little tobacco, mixing with it a plant which renders the fume less offensive. It is a social luxury, for the enjoyment of
which, they form a circle, and only one pipe is used. The principal
chief begins by drawing three whiffs, the first of which he sends up-
ward, and then passes the pipe to the person next in dignity, and in like
manner the instrument passes round until it comes to the first chief
again. He then draws four whiffs, the last of which he blows through
his nose, in two columns, in circling ascent, as through a double flued
chimney; and their pipes are not of the race stigmatized by
Knickerbocker as plebeian. None of the smoke of those villanous short
pipes, continually ascending in a cloud about the nose, penetrating
into and befogging the cerebellum, drying up all the kindly moisture of
the brain, and rendering the people who use them vapourish and testy;
or, what is worse, from being goodly, burly, sleek-conditioned men, to
become like the Dutch yeomanry who smoked short pipes, a lantern-
jawed, smoke-dried, leathern-hided race. The red people, whether of
the Rocky Mountains or of the Mississippi, belonged to the aristocracy
of the *long pipes*. Let us hope that they have not degenerated, and become followers of the costumes of the barbarian *ultra-marines*.

Turn over the leaves of *Westward Ho!* until you reach the end of
the seventh chapter, and then read of Salvation Yeo and his fiery reputation, and his eulogium—"for when all things were made, none was
made better than this; to be a lone man's companion, a bachelor's
friend, a hungry man's food, a sad man's cordial, a wakeful man's sleep,
and a chilly man's fire, sir; while, for stanching of wounds, purging of
rheum, and settling of the stomach, there's no herb like unto it under
the canopy of heaven." The truth of which eulogium Amyas testeth
in after years. But, "mark in the meanwhile," says one of the veracious
chroniclers from whom I draw these facts, writing seemingly in the
palmy days of good Queen Anne and

> not having [as he says] before his eyes the fear of that misocapnic
> Solomon James I or of any other lying Stuart, that not to South
> Devon, but to North; not to Sir Walter Raleigh, but to Sir Amyas

Leigh; not to the banks of the Dart, but to the banks of Torridge, does Europe owe the dayspring of the latter age, that age of smoke which shall endure and thrive when the age of brass shall have vanished, like those of iron and of gold, for whereas Mr. Lane is said to have brought home that divine weed (as Spenser well names it), from Virginia, in the year 1584, it is hereby indisputable that full four years earlier, by the bridge of Pulford in the Torridge moors (which all true smokers shall hereafter visit as a hallowed spot and point of pilgrimage) first twinkled that fiery beacon and beneficent loadstar of Bidefordian commerce, to spread hereafter from port to port, and peak to peak like the watch-fires which proclaimed the coming of the Armada and the fall of Troy, even to the shores of the Bosphorus, the peaks of the Caucasus, and the farthest isles of the Malayan sea; while Bideford, metropolis of tobacco, saw her Pool choked up with Virginian traders, and the pavement of her Bridgeland Street groaning beneath the savoury bales of roll Trinidado, leaf, and pudding; and the grave burghers, bolstered and blocked out of their own houses by the scarce less savoury stockfish casks which filled cellar, parlour, and attic, were fain to sit outside the door, a silver pipe in every strong right hand, and each left hand chinking cheerfully the doubloons deep lodged in the auriferous caverns of their trunkhose; while in those fairy rings of fragrant mist, which circled round their contemplative brows, flitted most pleasant visions of Wiltshire farmers jogging into Sherborne fair, their heaviest shillings in their pockets to buy (unless old Aubrey lies) the lotus leaf of Torridge for its weight in silver, and draw from thence, after the example of the Caciques of Dariena, supplies of inspiration much needed then, as now, in those Gothamite regions. And yet did these improve, as Englishmen, upon the method of those heathen savages; for the latter (so Salvation Yeo reported as a truth, and Dampier's surgeon, Mr. Wafer, after him), when they will deliberate of war or policy, sit round in the hut of the chief; where being placed, enter to them a small boy with a cigarro of the bigness of a rolling pin, and puffs the smoke thereof into the face of each warrior, from the eldest to the youngest; while they, putting their hand funnel-wise round their mouths, draw into the sinuosities of the brain, that more than Delphic vapour of prophecy; which boy presently falls down in a swoon, and being dragged out by the heels

and laid by to sober, enter another to puff at the sacred cigarro, till
he is dragged out likewise, and so on till the tobacco is finished, and
the seed of wisdom has sprouted in every soul into the tree of medi-
tation, bearing the flowers of eloquence, and, in due time, the fruit
of valiant action.

And with this quaint fact, narrated in the bombastic style of chronicles,
closeth the seventh chapter of the voyages and adventures of Sir Amyas
Leigh, under the style and title already mentioned, and after which
digression the course of our narrative proceedeth as before.

The inhabitants of Yemen smoke their well-loved dschihschi pipes,
with long stems passed through water, that the smoke may come cold
to the mouth; and which, when a few inveterate smokers meet together, keep up a boiling and bubbling noise, not unlike a distant corps
of drummers in full performance.

In the Austrian dominions, the lovers of the pipe may be found
amongst all classes of the community. Köhl writes, that after taking
two or three pipes of tobacco with the pasha at New Orsova, he went
into the market-place, where he found several merchants who invited
him to sit down, and again he was presented with a pipe. From this
place he went to a mosque, calling in at a school on his way: "The
little Turkish students were making a most heathenish noise, which
contrasted amusingly with the quiet and sedate demeanour of their
teacher, who lay stretched upon a bench, where he smoked his pipe,
and said nothing." He afterwards went to look at the fortifications,
and here and there saw a sentinel, with his musket in one hand and
pipe in the other. "Twenty-five soldiers were seen smoking under a
shed, and on the ground lay a number of shells or hollow balls, which
they assured us were filled with powder and other combustibles, yet
the soldiers smoked among them unconcernedly, and allowed us to do
the same." A gentleman from Constantinople told him that he had
seen worse instances of carelessness, in Asia Minor. He had there been
one day in the tents of a pasha, where some wet powder was drying
and being made into cartridges, and the men engaged in the work
were smoking all the while.

In the *Stettin Gazette* lately appeared a notification that the Prus-
sian clergy had privately been requested by the higher authorities to
abstain from smoking in public. We are not accustomed to it, and

should certainly think it odd to see clergymen perambulating the streets with short pipes in their mouths.

In all parts of the Sultan's dominions, the pipe or narghilè has a stem generally flexible, about six feet in length; and at this the owner will suck for hours. You may see a man travelling, mounted aloft on a tall camel, with his body oscillating to a and fro like a sailor's when he rows, but still that man has two yards of pipe before him. You may see two men caulking a ship's side as she lies careened near the shore. Up to their waists in water, they act up to the principle of division of labour; for one will smoke as the other plies the hammer, and then the worker takes his turn at the narghilè. Arabs sitting at work, fix their pipes in the sand. In the potteries both hands must be employed— how, then, can the potter smoke? Necessity is the mother of invention. One end of the pipe is suspended by a cord from the ceiling, the other is in the potter's mouth.

In smoking, Lane informs us, the people of Egypt and other countries of the East draw in their breath freely, so that much of the smoke descends into the lungs; and the terms which they use to express "smoking tobacco," signify "*drinking* smoke," or "*drinking tobacco*"; for the same word signifies both smoke and tobacco. Few of them spit while smoking; he had seldom seen them do so.

It was something like drinking of smoke that Napoleon accomplished in his unsuccessful smoking campaign. He once took a fancy to try to smoke. Everything was prepared for him, and his Majesty took the amber mouth-piece of the narghilè between his lips; he contented himself with opening and shutting his mouth alternately, without in the least drawing his breath. "The devil," he replied, "why, there's no result!" It was shewn that he made the attempt badly, and the proper method practically exhibited to him. At last he drew in a mouthful, when the smoke—which he had discovered the means of drawing in, but knew not how to expel—found its way into his throat, and thence by his nose, almost blinding him. As soon as he recovered breath, he cried out—"Away with it! What an abomination! Oh! the hog—my stomach turns!" In fact, the annoyance continued for an hour, and he renounced for ever a habit which, he said, was fit only to amuse sluggards.

Although Napoleon managed to fail, thousands less mighty have managed to succeed. There is a curious kind of legend mentioned in

Brand's *Antiquities*, by way of accounting for the frequent use and con-
tinuance of taking tobacco, for the veracity of which he declares that
he will not vouch. "When the Christians first discovered America,
the devil was afraid of losing his hold of the people there by the appearance of Christianity. He is reported to have told some Indians of
his acquaintance, that he had found a way to be revenged on the
Christians for beating up his quarters, for he would teach them to take
tobacco, to which, when they had once tasted it, they should become
perpetual slaves.

Without venturing to authenticate this strange story, in the moral
of which Napoleon would have concurred—with a mental reserva-
tion in favour of snuff—after the above defeat, let us console tobacco
lovers, that whilst the success of the first temptation closed the gates
of Paradise, the success of the second opens them again.

The following from an old collection of epigrams is, in every respect, worthy of the theme.

> All dainty meats I do defie,
> Which feed men fat as swine;
> He is a frugal man indeed
> That on a leaf can dine.
> He needs no napkin for his hands
> His fingers' ends to wipe,
> That keeps his kitchen in a box,
> And roast meat in a pipe.

In Hamburg, 40,000 cigars are smoked daily in a population
scarcely amounting to 45,000 adult males. And in London, the consumption must be considerable to furnish, from the profits of retail-
ing, a living to 1566 tobacconists. In England, we may presume that
the largest smoker of tobacco must be the Queen, since an immense
kiln at the docks, called the Queen's pipe, is occasionally lighted and
primed with hundredweights of tobacco, sea damaged or otherwise
spoiled, at the same time blowing a cloud

Which Turks might envy, Africans adore.

The total number of cigars consumed in France in 1857 is stated to
have been 523,636,000; and the total revenue of the French Government from the tobacco monopoly is estimated at £7,320,000 annually.

In Russia the revenue is £7,200,000 annually; and in Austria near £3,000,000. These are large sums to pay for the privilege of puffing.

The *Buffalo Democracy* estimates the annual consumption of tobacco at 4,000,000,000 of pounds. This is all smoked, chewed, or snuffed. Suppose it was all made into cigars 100 to the pound, it would produce 400,000,000,000 of cigars. These cigars, at the usual length, four inches, if joined together, would form one continuous cigar 25,253,520 miles long, which would encircle the earth more than 1000 times. Cut up into equal pieces, 250,000 miles in length, there would be over 1000 cigars which would extend from the centre of the earth to the centre of the moon. Put these cigars into boxes 10 inches long, 4 inches wide, and 3 inches high, 100 to the box, and it would require 4,000,000,000 boxes to contain them. Pile up these boxes in a solid mass, and they would occupy a space of 294,444,444 cubic feet; if piled up 20 feet high, they would cover a farm of 338 acres; and if laid side by side, the boxes would cover nearly 20,000 acres. Allowing this tobacco, in its unmanufactured state, to cost sixpence a pound, and we have 100,000,000 pounds sterling expended yearly upon this weed; at least one-and-a-half times as much more is required to manufacture it into a marketable form, and dispose of it to the consumer. At the very lowest estimate, then, the human family expend every year £250,000,000 in the gratification of an acquired habit, or a crown for every man, woman, and child upon the earth. This sum, the writer calculates, would build 2 railroads round the earth at a cost of £5,000 per mile, or 16 railroads from the Atlantic to the Pacific. It would build 100,000 churches, costing £2,500 each, or 1,000,000 dwellings costing £25 each (rather small!) It would employ 1,000,000 of preachers and 1,000,000 of teachers, giving each a salary of £125. It would support $3^1/_3$ millions of young men at college, allowing to each £75 a year for expenses.

What a cloud the "human family" would blow if they had each his share of the 4,000,000,000 pounds dealt out to him in cigars on the morning of the 25th of December, in the year of our Lord, 1860. One feels dubious as to the number who would refuse to take their quota, if there were nothing to pay.

Dr. Dwight Baldwin states, that in 1851, the city of New York spent 3,650,000 dollars for cigars alone, while it only spent 3,102,500 dollars for bread. The Grand Erie Canal, 364 miles long, the longest in the world, with its eighteen aqueducts, and eighty-four locks, was made in six years, at a cost of 7,000,000 dollars. The cigar bill in the city of New York would have paid the whole in two years.

The number of cigar manufactories in America is 1,400, and the number of hands employed in them 7,000 and upwards. The total estimated weekly produce of these manufactories is 17 millions, and the yearly 840 millions. At 7 dollars per 1,000, these would be worth 5 million dollars, and adding 50 per cent. for jobber and retailer, the total cost to consumers would be $7^1/_2$ million dollars—add to this the sum paid for imported cigars, 6 million dollars, and we have $13^1/_2$ million dollars, the value of cigars consumed yearly in the United States, without adding profit to the imported cigars; so that, including the amount expended in tobacco for smoking and chewing, and in snuff, the annual cost of the tobacco consumed yearly, is not less than 30 million dollars or £6,000,000. This is but little more than is realized annually in Great Britain by the excise duty alone on the tobacco consumed at home; but it must be remembered, that in America tobacco is free of the duty of three shillings and twopence per pound, and free of charges for an Atlantic passage, so that the tobacco represented by 6 millions there, would be represented here by at least six times that amount.

Cloudland costs something to keep up its dignity after all, but beauty is seductive, and so is tobacco.

Yes! St. John (Percy, we mean—not "the Divine"), there must be "magic in the cigar." Then, to the sailor, on the wide and tossing ocean, what consolation is there, save in his old pipe! While smoking his inch and a half of clay, black and polished, his Susan or his Mary becomes manifest before him, he sees her, holds converse with her spirit—in the red glare from the ebony bowl, as he walks the deck at night, or squats on the windlass, are reflected the bright sparkling eyes of his sweetheart. The Irish fruit-woman, the Jarvie without a fare, the policeman on a quiet beat, the soldier at his ease, all bow to the mystic power of tobacco*—all acknowledge the infatuations of *Cloudland*.

Bentley's Magazine.

CHAPTER V
PIPEOLOGY

It was his constant companion and solace. Was he gay, he smoked—was he sad, he smoked—his pipe was never out of his mouth—it was a part of his physiognomy; without it his best friends would not know him. Take away his pipe—you might as well take away his nose.

Knickerbocker's *New York*

Semele, in a death by fire, became a martyr to love. Thus Virginia suffers herself to be burnt for the good of the world. From the ashes of the old Phœnix the young Phœnix was born. From the smoke of the Havana spring new visions, and eloquent delights. As the altars of the gods received honour from men, and the censers from whence ascended the burning incense were sacred to the deities, wherefore should not the pipe receive honour, as well as the man who uses it, or the odorous weed consumed within it. An enthusiast writes of it thus—

Philosophers have drawn their best similes from their pipes. How could they have done so, had their pipes first been drawn from them? We see the smoke go upwards—we think of life; we see the smoke-wreath fade away—we remember the morning cloud. Our pipe breaks—we mourn the fragility of earthly pleasures. We smoke it to an end, and tapping out the ashes, remember that "Dust we are, and unto dust we shall return." If we are in love, we garnish a whole sonnet with images drawn from smoking, and first fill our pipe, and then tune it. That spark kindles like her eye, is ruddy as her lip; this slender clay, as white as her hand, and slim as her waist; till her

45

raven hair grows grey as these ashes, I will love her. This perfume is
not sweeter than her breath, though sweeter than all else. The odour
ascends me into the brain, fills it full of all fiery delectable shapes,
which delivered over to the tongue, which is the birth become delectable wit.

The instruments by which the "universal weed" is consumed, are
almost as variable in form and material as the nations indulging in
their use. The pipe of Holland is of porcelain, and that of our own
island of unglazed clay. These latter are made in large quantities, both
at home and abroad.* One factory at St. Omer employs 450 work-
people, and produces 200,000 gross, or nearly thirty millions of pipes,
consuming nearly eight thousand tons of clay in their manufacture.
The quantity of pipes used annually in London is estimated at 354,000
gross, or 52,416,000 pipes; it requires 300 men, each man making 20
gross four dozen per week, for one year, to make them; the cost of
which is £40,950. The average length of these pipes is twelve and a
half inches; and if laid down in a horizontal position, end to end together, they would reach to the extent of 10,340 miles, 1,600 yards; if
they were piled one above another perpendicularly, they would reach
135,138 times as high as St. Pauls; they would weigh 1,137 tons, 10
cwts., and it would require 104 tons, 9 cwts., 32 pounds of tobacco to
fill them. In 1857 we imported clay pipes to the value of £7,614, which
cannot be short of 121,000 gross, or seventeen and a half millions.
But even with us, pipes were not always of clay. The earliest pipes
used in Britain are stated to have been made from a walnut-shell and
a straw. Dr. Royle describes a very primitive kind of clay pipe used by
some of the natives of India—it is presumed only in cases of necessity.
"The amateur makes two holes, one longer than the other, with a
piece of stick in a clay soil, inclining the stick so that they may meet;
into the shorter hole he places the tobacco, and applies his mouth to
the other, and thus, as he lies upon the ground, luxuriates in the fumes
of the narcotic herb."

---

*For the art of making tobacco-pipes of clay, the Dutch are indebted to this
country, in proof of which, Mr. Hollis, who passed through the Netherlands
in 1748, states that the master of the Gouda Pipe Works informed him, that,
to that day, the principal working tools bore English names.

Turkish pipe-bowls, or Lules, are composed of the red clay of Nish, mixed with the white earth of the Roustchouck. They are very graceful in form, and are in some cases ornamented with gilding. The "regular Turk" prefers a fresh bowl daily; therefore the plain ones are resorted to on the score of economy. In Turkey and some other parts of the Orient, it is not unusual to compute distances, or rather the duration of a journey, by the numbers of pipes which might be smoked in the time necessary to accomplish it.

The pipe of the German is, almost universally, the Meerschaum, that pipe of fame so coveted by the Northern smoker. These articles are composed of a kind of magnesian earth, known to the Tartars of the Crimea as *keff-til*. Pallas erroneously supposed that this kind of earth was so denominated from Caffa, and therefore the name signified "Caffa earth." From *Meninshi's Oriental Dictionary* it would appear to be a derivation of two Turkish words which signify "foam" or "froth" of the "earth." The French name, *ecume de mer*, or "scum of the sea," and the Germans "sea foam," have doubtless an intimate relationship with this same keff-til of the Crimean Tartars.

Meerschaum earth is met with in various localities in Spain, Greece, Crimea, and Moravia. The greatest quantity is derived from Asia Minor, it being dug principally in the peninsula of Natolia, near the town of Coniah. Before the capture of the Crimea, this earth is stated to have formed a considerable article of commerce with Constantinople, where it was used in the public baths to cleanse the hair of women. The first rude shape was formerly given to the pipe-bowls on the spot where the mineral was dug, by pressure in a mould; and these rude bowls were more elegantly carved and finished at Pesth and Vienna. At the present time, the greater part of the meerschaum is exported in the shape of irregular blocks; these undergo a careful manipulation, after having been soaked in a preparation of wax and oil. After being finished, and sold at the German fairs, some of them have acquired such an exquisite tint through smoking, in the estimation of connoisseurs, that they have realized from £40 to £50.

Attempts have not been wanting to imitate this material, hitherto not very successfully. The large quantity of paring that are left in trimming up the bowls, has been rendered available for the manufacture of what are called "massa bowls," but they do not enjoy the reputation of the genuine meerschaum bowls.

There is yet another mineral production, the use of which Turkish smokers, at any rate, know how to appreciate. This is amber. The Turk will expend an almost fabulous sum in an amber mouthpiece for his narghilè. Four valuable articles of this description were exhibited in the Turkish department of the Exhibition of 1851, which were worth together £1000, two of them being valued at £305 each. There is a current belief in Turkey that amber is incapable of transmitting infection; and as it is considered a great mark of politeness to offer the pipe to a stranger, this presumed property of amber accounts in some measure for the estimation in which it is held.

The knowledge of amber extends backwards to a remote antiquity, as the Phœnicians of old fetched it from Prussia. Since that period it has been obtained there uninterruptedly, without any diminution in the quantity annually collected. The greatest amount of amber is found on the coast of Prussia proper, between Konigsberg and Dantzic. From the amber-beds on the coast of Dirschkeim, extending under the sea, a storm threw up, on the 1st of January, 1848, no less than 800 pounds. The amber fishery of Prussia formerly produced to the king about 25,000 crowns per month. After a storm, the amber coasts are crowded with gatherers, large masses of amber being occasionally cast up by the waves. In digging for a well in the coal-mines near Prague, the workmen lately discovered, between the bed of gritstone which forms the roof of that mine and the first layer of coals, a bed of yellow amber, apparently of great extent. Pieces weighing from two to three pounds have been extracted. There are two kinds— the terrestrial, which is dug in mines, and the marine, which is cast ashore during autumnal storms.

Opinions vary as to the origin of amber. Tacitus and others have considered it a fossil resin exhaled by certain coniferous trees, traces of which are frequently observed among the amber, whilst other theorists contend that it is a species of wax or fat, having undergone a slow process of putrefaction; this latter view being based upon the fact that chemists are able to covert fatty or cerous substances into succinic acid by artificial oxidation. One thing is, however, certain, that amber, at some period of history, must have existed in a state of fluidity, since numerous insects, especially of the spider kind, are found imbedded in it; and a specimen has been shown enclosing the leg of a toad. Toads are in the habit of living for centuries, we are informed, cooped up in stone and rock; but we are not aware that hitherto any of these extraordinary reptiles have been found buried alive in a mass of amber. Masses of amber have been found weighing from 4 to 6 pounds—more than large enough to contain a toad or two of ordinary dimensions.

For a knowledge of the pipes of modern Egypt, we must resort for information to Mr. Lane, from whom we gather the following notes. The pipe (which is called by many names, as "shibuk," "ood," etc.) is generally between four and five feet long. Some pipes are shorter, and some of greater length. The most common kind used in Egypt is made of a kind of wood called "garmashak." The greater part of the stick is covered with silk, which is confined at each extremity by gold thread, often intertwined with coloured silks, or by a tube of gilt silver; and at the lower extremity of the covering is a tassel of silk. The covering was originally designed to be moistened with water, in order to cool the pipe, and consequently the smoke, by evaporation; but this is only done when the pipe is old, or not handsome. Cherrystick pipes, which are never covered, are used by some persons, particulary in the winter. In summer, the smoke is not so cool from the cherrystick pipe as from the kind before mentioned. The bowl is of baked earth, coloured red or brown. The mouthpiece is composed of two or more pieces of opaque, light-coloured amber, interjoined by ornaments of enamelled gold, agate, jasper, carnelion, or some other precious substance. This is the most costly part of the pipe. Those in ordinary use by persons of the middle classes cost from £1 to £3 sterling. A wooden tube passes through it; this is often changed, as it becomes foul from the oil of the tobacco. The pipe also requires to be cleaned very often, which is done with tow, by means of a long wire. Many poor men in Cairo gain a livelihood by cleaning pipes. Some of the Egyptians use the Persian pipe, in which the smoke passes through water. The pipe of this kind most commonly used by persons of the higher classes is called narghilè, because the vessel that contains the water is the shell of a cocoa-nut, of which narghilè is an Arabic name. Another kind which has a glass vase, is called "sheesheh," from the Persian word signifying "glass." Each has a very long, flexible tube.

A kind of pipe commonly called "gozeh," which is similar to the narghilè, excepting that it has a short cane tube, instead of the snake, and no stand. This is used by men of the lowest class for smoking both the "tumbak" or Persian tobacco, and the narcotic hemp.

The Zoolus of Southern Africa have a kind of pipe or smoking horn called "Egoodu," which is constructed on a similar principle to the Persian pipe. The herb is placed at the end of a reed introduced into the side of an oxhorn, which is filled with water, and the mouth applied to the upper or wide part of the horn, the smoke passing down the reed and through the water.

The Delagoans of Eastern Africa smoke the "hubble-bubble," a similar instrument, having the upper part of the horn closed, excepting a small orifice in the centre of the covering through which the smoke is inhaled.

The Kaffirs form pipe bowls from a black, and also from a green, stone; they are in shape similar to the Dutch pipes, and without ornament. The negroes of Western Africa have pipes of a reddish earth, some of them of very uncouth and singular forms, others close imitations of European pipe bowls. One kind of pipe consists of two bowls placed side by side upon a single stem. Old Indian pipes have been found in America, also fashioned out of green stone.

The natives of the South-West coast of Africa, near Elizabeth's Bay, use pipes in the shape of a cigar tube formed of a mottled green or white mineral of the magnesian family, externally carved or roughly ornamented.

Sailors, when on a voyage, are often in difficulties for the want of pipes. Under such circumstances, numerous contrivances have at different times been resorted to remedy the defect; such as pipes cast out of old lead, or cut out of wood. The sailors belonging to H.M.S. *Samarang* having lost their pipes in the Sarawak river, set to, and in a very little while, manufactured excellent pipes from different sized internodes of the bamboos that grew around them. In India, simple pipes are used composed of two pieces of bamboo, one for the bowl cut close to a knot, and a smaller one for the tube.

The aborigines of British Guiana use a pipe, or rather a tube, called a "Winna." It resembles a cheroot in outward appearance, but is hollow, so as to contain the tobacco. It is said to be made from the rind of the fruit of the manicot palm, growing on the river Berbice. Forasmuch as it pleaseth us to borrow fashions from nations barbarous as well as civilized, a form of tube much resembling the Winna, has been made and sold in the tobacconist shops of the metropolis of old England.

Among the Bashee group, and particularly on the island of Ibayat, the natives form very elegant and commodious pipes from different species of shells, the columella and septa of the convolutions being broken down, and a short ebony stem inserted into a hole at the apex of the spire. These are more generally formed of the shells known as the Bishop's mitre (*Mitra episcopalis*) and the Pope's mitre (*Mitra papalis*). Species of *Terebra* and *Turbo* are also converted into pipes.

In China, where M. Rondot calculates that there are not less than 100 millions, and Abbè Huc 300 millions of smokers, pipes are made in immense numbers. Of these there are three kinds, the water pipe, the straight pipe, and the opium pipe. Chinese pipes, and indeed those of all the Indo-Chinese races, including the Tartars, Chinese, Koreans, and Japanese, are provided with a small metallic bowl, and usually a long bamboo stem; for with persons who are in the habit of smoking, at short intervals, all day long, a large bowl would be inadmissable. By inhaling but a pinch of tobacco on one occasion, they extend the influence of a larger pipe over a greater space of time. In such cases they suffer no inconvenience from the nature of the material of which the bowl is composed. Nations that smoke larger pipes adopt some other substance, as metal would become too hot; hence we have pipes of Samian ware in Turkey, Meerschaum in Germany, and clay in England and other places. My Uncle Toby would have burnt his fingers with a Chinese pipe of nickel silver many a time and often; and it would have required a large amount of logic to have induced Doctor Riccabocco to have exchanged his companion (his pipe, not his umbrella) for a bowl of Japanese manufacture.

Isaac Browne thought, a century ago, that there was something in a pipe worth writing about, or he would have never given us the following:

Ode to a Tobacco Pipe

Little tube of mighty power,
Charmer of an idle hour,
Object of my warm desire,
Lip of wax, and eye of fire;
And thy snowy taper waist,
With thy finger gently braced;
And thy pretty swelling crest,

With thy little stopper prest;
And the sweetest bliss of blisses
Breathing from thy balmy kisses.
Happy thrice, and thrice again,
Happiest he of happy men;
Who, when again the night returns,
When again the taper burns,
When again the cricket's gay
(Little cricket full of play),
Can afford his tube to feed
With the fragrant Indian weed;
Pleasure for a nose divine,
Incense of the god of wine.
Happy thrice, and thrice again,
Happiest he of happy men.

In Virginia's native country, the pipe sticks closer to a man than his boots. An American is no more furnished without his pipe or ci-gar, than a house is furnished without a looking glass. To the native Indian, it supplies an important place; it becomes his treaty of peace—his challenge of war. It is the instrument of a solemn ratification, and the subject of more than one semi-sacred legend, which has woven about the heart of the Red-man.

"At the Red-pipe Stone Quarry," say they,

happened the mysterious birth of the red-pipe, which has blown its fumes of peace or war to the remotest corners of the Continent, which has visited every warrior, and passed through its reddened stem, the irrevocable oath of war and desolation. And here, also, the peace breathing calumet was born, and fringed with the eagle's quills, which has shed its thrilling fumes over the land, and soothed the fury of the relentless savage. The Great Spirit, at an ancient period, here called together the Indian warriors, and standing on the precipice of the red-pipe stone rock, broke from its wall a piece, and made a huge pipe, by turning it in his hand, which he smoked over them, and to the north, the south, the east, and the west; and told them that this stone was red—that it was their flesh—that they must use it for their pipes of peace, that it belonged to them all, and that the war club, and the scalping knife must not be raised on its ground. At the last whiff of his pipe, his head went into a great cloud, and the whole surface of the rock, for several miles, was melted and glazed. Two great ovens were opened beneath, and two women, guardian spirits of the place, entered them in a blaze of fire, and they are heard there yet, answering to the invocations of the priests or medicine men, who consult them when they are visitors to this sacred place.*

From the red stone of the quarry
With his hand he broke a fragment,
Moulded it into a pipe head,
Shaped and fashioned it with figures.
From the margin of the river
Took a long reed for a pipe stem,
With its dark green leaves upon it;
Filled the pipe with bark of willow;
With the bark of the red willow;
Breathed upon the neighbouring forest,
Made its great boughs chafe together,
Till in flame they burst, and kindled;
And erect upon the mountains,
Gitche Manito, the mighty,
Smoked the calumet, the Peace Pipe,
As a signal to the nations. . . .

The tribes of the Missouri make their pipes of a stone called Catlinite, from the red pipe stone quarries upon the head waters of that river, the colour of which is brick red. These stones, when first taken out of the quarry are soft, and easily worked with a knife, but on exposure to the air become hard and take a good polish. The pipes of the Rocky Mountain Indians are some of them wrought with much labour and ingenuity of an argillaceous stone of a very fine texture, found at the north of Queen Charlotte's Island. This stone is of a blue black colour, and in character similar to the red earth of the Missouri quarry.

---

*Catlin's *North American Indians*, vol. ii., pp. 160.

The calumet or "pipe of peace" of the Sioux Indians is thus described by Irving. "The bowl was a species of red stone resembling porphyry, the stem was six feet in length, decorated with tufts of horse hair dyed red. The pipe bearer stepped within the circle, lighted the pipe, held it towards the sun, then towards the different points of the compass, after which he handed it to the principal chief. The latter smoked a few whiffs, then, holding the head of the pipe in his hand, offered the other end to their visitor, and to each one successively in the circle. When all had smoked, it was considered that an assurance of good faith and amity had been interchanged." The use of the uspogan or calumet among the Eythinyuwak, appears not to have been an original practice of the Tinne, but was introduced with tobacco by Europeans; while among the Chippeways, the plant has been grown from the most ancient times.

Among the most uncultivated and uncivilized of nations, the pipe is an object upon which is exercised all their ingenuity, and in the decoration of which is concentrated all their taste. One might almost classify the races of the world by means of a good collection of their pipes, and not stray very far from the order resulting from more scientific processes.

In the East, there is existing an almost incessant habit of smoking; and the pipe is the prelude of all official acts, of all conversations, and of all social relations. The Oriental seizes his pipe in the morning, and scarcely relinquishes it till he goes to bed. Here there is generally a special functionary—the pipe-bearer—as an appendage to all officials. When the Sultan goes abroad, his pipe-bearer is with him. In families of respectability, the care of the pipes is the exclusive attribute of one or more servants, who occupy the highest grade of the domestic establishment; and thus dignity is given to the pipe, even in a country where less dignity is allowed to the fairer portion of the community than in more highly cultivated countries.

In the Museum of the Botanic Gardens at Kew, are pipes and stems carved out of boxwood, as used in Sweden; also pipe-bowls of pine and other woods made of the native Indians near Sitka in North-West America, and brought home from a late expedition. The latter are rude, but quite equal in elegance to many which adorn the windows of fancy tobacconists and cigar divans in this metropolis of the civilized world.

From a schism in tobacco-pipes, Knickerbocker dates the rise of parties in the *Niew Nederlandts*. "The rich and self-important burghers, who had made their fortunes, and could afford to be lazy, adhered to the ancient fashion, and formed a kind of aristocracy, known as the *Long-pipes*; while the lower order, adopting the reform of William Kieft, as more convenient in their handicraft employments, were branded with the plebeian name of *Short-pipes*." Who may be considered as the founder of the English Short-pipe school, is more difficult to determine; it is nevertheless, of late years, a very popular one, and considerably out-numbers the aristocracy of Long-pipes. The variety of these instruments is almost infinite. There are all kinds of short clays, cutties, St. Omer, Gambier, meerschaum washed, coloured clay, and fancy clay of all shapes, grotesque, uncouth, stupid, and in some instances graceful. Pipes also of wood, of black ebony, green ebony, brier-root—whatever that may be—cherry-root, tulip-wood, rosewood, etc. Glass pipes, with reservoirs and without, smokers' friends, and, if we may judge from their size, tobacconists' friends; meerschaum bowls, massa bowls, porcelain bowls, clay bowls, or uncouth and monstrous heads, with eyes of glass and enamelled teeth, together with short stems and mounts for broken clays. Add to these, one knows not how many kinds of tobacco-pots, from a smiling damsel in all the glories of crinoline, to the dissevered head of Poor Dog Tray. The windows of retail tobacconists now-a-days more resemble a toy-shop, or a fancy stall from an arcade or bazaar, than the sober-looking windows of a retailer half a century ago. Mr. Frank Fowler informs us that the same tastes have migrated to Australia.

> The cutty is of all shapes, sizes, and shades. Some are negro heads, set with rows of very white teeth; some are mermaids, showing their more presentable halves up the front of the bowls, and stowing away their weedy extremities under the stems. Some are Turkish caps, some are Russian skulls, some are houris, some are Empresses of the French, some are Margaret Catchpoles, some are as small as my lady's thimble, others as large as an old Chelsea tea-cup. Everybody has one, from the little pinafore schoolboy, who has renounced his hardbake for his Hardham's, to the old veteran who came out with the second batch of convicts, and remembers George Barrington's prologue. Clergymen get up their sermons over the pipe; members

of parliament walk the verandah of the Sydney House of Legisla-
ture, with the black bowl gleaming between their teeth. One of the
metropolitan representatives was seriously ill just before I left, from
having smoked forty pipes of Latakia at one sitting. A cutty bowl,
like a Creole's eye, is most prized when blackest. Some smokers wrap
the bowls reverently in leather during the process of colouring; others
buy them ready stained, and get (I suppose) the reputation of accomplished whiffers at once. Every young swell glories in his cabinet of dirty clay pipes. A friend of mine used to call a box of the
little black things his "*Stowe* collection." Tobacco, I should add here,
is seldom sold in a cut form; each man carries a cake about with
him, like a card-case; each boy has his stick of Cavendish, like so
much candy. The cigars usually smoked are Manillas, which are as
cheap and good as can be met with in any part of the world. Lola
Montez, during her Australian tour, spoke well of them. What stronger puff could they have than hers?

CHAPTER VI

SNIFFING AND SNEESHIN

> "'Tis most excellent," said the monk. "Then do me the
> favour," I replied, "to accept of the box and all; and when
> you take a pinch out of it, sometimes recollect that it was the
> peace-offering of a man who once used you unkindly, but not
> from the heart."
>
> Sterne's *Sentimental Journey*

Everybody, of course, knows all about the Franciscan and his snuff-box, with which this chapter begins. Sterne narrates it in his happiest vein, and all who read it are somehow sure to remember it. Boxes are exchanged; the traveller is left to himself. Now he moralises:

> I guard this box as I would the instrumental parts of my religion, to
> help my mind on to something better. In truth, I seldom go abroad
> without it; and oft and many a time have I called up by it the cour-
> teous spirit of its owner to regulate my own in the justlings of the
> world. They had found full employment for his, as I learned from
> his story, till about the forty-fifth year of his age, when, upon some
> military services ill-requited, and meeting at the same time with a
> disappointment in the tenderest of passions, he abandoned the sword
> and the sex together, and took sanctuary, not so much in his con-
> vent as in himself.

The word "snuff" is stated by competent authorities, to be an in-
flection of the old northern verb *sniff*, which latter word was in existence long before the invention or knowledge of the substance to which

it now gives its name.* In its earlier signification, it was expressive of strong inhalation through the nostrils, or descriptive of any impatience. Hence arose the expressions in use in the sixteenth and seventeenth centuries, to "snuff pepper" or "take in snuff." Shakespeare makes a similar use of the phrase in *Henry IV*, in connection with a small box of perfume displayed by a courtier to the annoyance of Hotspur.

> *He was perfumed like a milliner;*
> *And, 'twixt his finger and his thumb, he held*
> *A pouncet box, which ever and anon*
> *He gave his nose, and took't away again;*
> *Who, therewith angry, when it next came there,*
> *Took it in snuff.*

In this quotation we also meet with the "pouncet box," which seems to have been a small box having a "pounced" or perforated cover, containing perfumes, the scent of which escaping from the open work at the top was regarded as a preservative against contagion. From the pouncet box the perfumes were inhaled. It was probably not till a century after the introduction of tobacco, that the triturated dust was commonly in use, and there became any occasion for the *snuff-box*.

Humboldt gives an account of a curious kind of snuff, as well as an extraordinary method of inhaling it, which came under his notice while travelling in South America. "The Ottomacs," he says,

> throw themselves into a peculiar state of intoxication, we might say of madness, by the use of the powder of *niopo*. They gather the long pods of an acacia (made known by him under the name of *Acacia niopo*), cut them into pieces, moisten them, and cause them to ferment. When the softened seeds begin to grow black, they are kneaded like paste, mixed with some flour of cassava and lime procured from the shell of a *helix* (snail), and the whole mass is exposed to a very brisk fire, on a gridiron made of hard wood. The hardened paste takes the form of small cakes. When it is to be used, it is reduced to a fine powder, and placed on a dish, five or six inches wide. The Ottomac holds this dish, which has a handle, in his right hand, while he inhales the niopo by the nose, through the forked bone of a bird, the two extremities of

---

*Tooke says "*snuff* is the past participle of to *sniff*, that which is *sniffed*."

which are applied to the nostrils. This bone, without which the Ottomac believes that he could not take this kind of snuff, is seven inches long; it appears to be the leg bone of a large species of plover. The niopo is so stimulating, that the smallest portions of it produce violent sneezing in those who are not accustomed to its use.

Father Gumilla says,

this diabolical powder of the Ottomacs, furnished by an aborescent tobacco plant, intoxicates them through the nostrils, deprives them of reason for some hours, and renders them furious in battle.

A custom analagous to this, La Condamine observed among the natives of the Upper Maranon. The Omaguas, a tribe whose name is intimately connected with the expeditions in search of El Dorado, have, like the Ottomacs, a dish, and the hollow bone of a bird, and a powder called *curupa*, which they convey to their nostrils by means of these, in a manner identical with that of the Ottomacs. This powder is also obtained from the seed of a kind of acacia, apparently closely allied to, if not the same as the niopo.

A similar instrument to the bone of the Ottomacs and Omaguas has already been referred to as in use in Hispaniola, for inhaling through the nostrils the smoke of burning tobacco leaves.

The method of taking snuff in Iceland is described by Madame Pfeiffer as differing from the methods above detailed, but equally singular. Most of the peasants, and many of the priests, have no proper snuff-box, but only a box made of bone, and shaped like a powder flask. When they take snuff, they throw back the head, insert the point of the flask in the nose, and shake a dose of snuff in it. They then offer it to their neighbour, who repeats the performance, passes it to his, and thus it goes the round, until it reaches its owner again. Had this been the custom in the days of the *Rape of the Lock*, Belinda had not so readily subdued the baron, as with one finger and a thumb—

Just where the breath of life his nostrils drew,
A charge of snuff the wily virgin threw;
The gnomes direct, to every atom just,
The pungent grains of titillating dust.
Sudden, with starting tears each eye o'erflows,
And the high dome re-echoes to his nose.

The Zoolus of Southern Africa use a small gourd to carry their snuff, and a small ivory spoon with which to ladle out the dust. We remember many years ago an elderly gentleman who practiced on the Zoolu plan, his snuff was carried loose in his waistcoat pocket, whence it was conveyed to his nose by means of a small silver spoon, which was always at hand for the purpose.

ZOOLU SNUFF GOURD AND SPOON

As early as the beginning of the reign of James I, a "taker of to-bacco" was furnished with an apparatus resembling that of a modern Scotch mull, when supplied with all the necessary implements. In 1609, Dekker, in his *Gull's Horn Book*, says—"Before the meat come smoking to the board, our gallant must draw out his tobacco-box, the ladle for the cold snuff into the nostril, the tongs and priming iron; all which artillery may be of gold or silver, if he can reach the price of it." In 1646, Howell describes the apparatus and practice of snuff taking as quite common in other countries; since, he says—"The Spaniards and Irish take tobacco most in powder or *smutchin*, and it mightily refreshes the brain; and I believe there's as much taken this way in Ireland, as there is in pipes in England. One shall commonly see the serving maid upon the washing block, and the swain upon the ploughshare, when they are tired of their labour, take out their boxes of smutchin, and draw it into their nostrils with a quill, and it will beget new spirits in them with a fresh vigour to fall to their work again."

The word printed "smutchin" by Howell, is stated to be more ac-curately "sneeshin," a vulgar name for snuff which causes sneezing; and hence "sneeshin mill" (sometimes corrupted into "mull") is the

Scottish name for snuff-box. *Dr. Jameson's Etymological Dictionary* may be considered as an authority in these matters; and from it we learn that the word "mill" is the vulgar name for a snuff-box, especially one of a cylindrical form, or resembling an inverted cone. No other name was formerly in use in Scotland; and the reason assigned for it is, that when tobacco was first introduced into this country, those who wished to have snuff, were accustomed to toast the tobacco leaves before the fire, and then bruise them with a piece of wood in the box, which was thence called a "mill," because the snuff was ground in it. From all this, it is easy to perceive how a ram's horn, from its conical shape, became one of the primitive forms of the Scottish snuff-box, although latterly it is often one of the most costly and luxurious.

In confirmation of the latter remark, it is only necessary to refer to an example in the Exhibition of 1851. Mr. W. Baird of Glasgow, exhibited a ram's head beautifully mounted, as a snuff-box and cigar case. When alive, he must have been a noble sheep, for the circular horns measured no less than 3 feet 4 inches from root to tip. The cigar case was beautifully mounted, having on the top a splendid Scotch amethyst, surmounted with thistle wreaths in gold and silver, and set out with many fine cairngorms and small amethysts. The snuff-box cavity, occupied the centre of the forehead, the lid surmounted by a splendid cairngorm, and clustered with gold and silver wreaths and small precious stones. In fact, the head presented a perfect flourish of the most beautiful and gracefully disposed ornaments, and altogether the article was most unique. Attached thereto was a fine ivory hammer and silver spoon, pricker and rake, with a silver mounted hare's foot. It ran on ivory castors upon a rosewood platform, surmounted by a glass shade. There were not less than nine hundred separate pieces of precious stones and metals used in the construction of this ornate article.

Down to the middle of the eighteenth century, the "sneeshin horn," with spoon and hare's foot attached to it by chains, appears to have been regarded as so completely a national characteristic, that when Baddeley played Gibby in *The Wonder*, with Garrick, he came on the stage with such an apparatus.

The Mongrabins and other African races, according to Werne, are much addicted to snuff taking. The snuff they usually carry in small oval-shaped cases made out of the fruit of the Doum palm; these have a very small opening at one end, stopped up by a wooden peg, and the snuff is not taken in pinches, but shaken out on the back of the hand. Mr. Campbell, while travelling in South Africa, gave a Bushman a piece of tobacco. It was speedily converted into snuff. One of the daughters, after grinding it between two stones, mixed it with white ashes from the fire; the mother then took a large pinch of the composition, putting the remainder into a piece of goat's skin, among the hair, and folding it up for future use.

The snuff in use in Africa is not always made from tobacco. Mr. Hutchinson states that he saw at Panda, on the western coast, snuff made of the powdered leaves of the monkey fruit tree (*Adansonia digitata*). That of the Zoolus is composed of the dried leaves of the dacca or narcotic hemp mixed with the powder of burnt aloes. Whether or not this was the kind of snuff which Mr. Richardson was knocked down with in his journey across the Great Desert, we are not in a position to determine; whatever it was, it appears to have been extremely powerful. "A merchant," he says, "offered me a pinch of snuff, and to please him I took a large pinch, pushing a portion of it up my nostrils. Immediately I fell dizzy and sick, and in a short time vomited violently. The people stared at me with astonishment, and were terrified out of their wits, and thought I was about to give up the ghost. They never saw snuff before produce such terrible effects. After some time I got a little better, and returned home. This snuff was from Souf, and is called *wâr* (difficult). I had been warned of it, and therefore paid richly for my folly; indeed, the Souf snuff is extremely powerful." Some of the strict Mahometans of Ghadames consider snuffing, as well as smoking, prohibited by their religion, and therefore do not indulge in it. The South American traveller which Mr. Lizars, the tobacco antagonist, once fell in with, was evidently not a strict Mahometan, for he first filled his nostrils with snuff, which he prevented falling out by stuffing shag tobacco after it, and this he termed "plugging"; then put in each cheek a coil of pig-tail tobacco, which he named "quidding"; lastly, he lit a Havannah cigar, which he put into his mouth, and thus smoked and chewed—puffing at one time the smoke of the cigar, and at another time squirting the juice from his mouth. What a phenomenon! That gentleman should have politely thanked the South American for permitting him to view an exhibition, such as he may never have the pleasure of seeing again. And what a capital illustration ready made to his hands. It is almost equal to those elaborate calculations which are based upon the amount of time consumed in taking so many pinches of snuff during the day, and so many repetitions of the operation of blowing the nose.*

A correspondent of the *Petersburg* (Va) *Express* says—

> There are, perhaps, in our state 125,000 women, leaving out of the account those who have not cut their teeth, and those who have lost them from age. Of this number, eighty per cent. may be safely set down as snuff-dippers. Every five of these will use a two-ounce paper of snuff per day—that is to the 100,000 dippers 2,500 lbs. a day, amounting to the enormous quantity of 912,000 lbs. In this number of snuff-dippers are included all ages, colours, and conditions. This practice is generally prevalent in the pine districts of North Carolina, and in many parts of South Carolina, Georgia, Alabama, Florida, and Eastern Tennessee. It may be thus described:—A female snuff-dipper takes a short stick, and, wetting it, dips it into her snuff-box, and then rubs the gathered dust all about her mouth, into the interstices of her teeth, &c., where she allows it to remain until its strength has been fully absorbed. Others hold the stick thus loaded with snuff in the cheek, *a la* quid of tobacco, and suck it with a decided relish, while engaged in their ordinary avocations; while others simply fill the mouth with the snuff, and thus imitate, to all intents and purposes, the chewing propensities of the men. In the absence of snuff, tobacco, in the plug or leaf, is invariably resorted to as a substitute. Oriental betel chewing is elegant, compared to "snuff-dipping."

---

*Lord Stanhope makes the following curious estimate: "Every professed, inveterate, and incurable snuff-taker, at a moderate computation, takes one pinch in ten minutes. Every pinch, with the agreeable ceremony of blowing and wiping the nose, and other incidental circumstances, consumes a minute and a half. One minute and a half out of every ten, allowing sixteen hours to a snuff-taking day, amounts to two hours and twenty minutes out of every natural day, or one day out of every ten. One day out of every ten amounts to thirty-six days and a half in the year; hence, if we suppose the practice to be persisted in for forty years, two entire years of the snuff-taker's life will be dedicated to tickling his nose, and two more to blowing it." The expense of snuff, snuff-boxes, and handkerchiefs, is also alluded to; and it is calculated that "by a proper application of the time and money thus lost to the public, a fund might be constituted for the discharge of the national debt."

The most uncomfortable reflection to the snuffer is that which concerns the probability of his consuming himself by a condition of slow poisoning, not the result of the pure tobacco, but its impure associates in the box. In boxes lined with very thin lead, but especially in cases where the leaden lining is thicker, and which are much used by the Paris retailers, a chemical action takes place, the result of which is to charge the snuff with subacetate of lead. This result was suspected by Chevalier, and has been confirmed by Boudet of Paris, and Mayer of Berlin, by careful experiments. Mayer traces several deaths and cases of saturnine paralysis to the patient's having taken snuff from packets, the inner envelope of which was thin sheet lead, in constant contact with the powdered weed. The cry once heard of "death in the pot," requires now to be exchanged for "death in the box," and Holbein to give us a new plate of the skeleton form emerging from a packet or snuff-box containing the scented rappee.

Late investigations have shown that no small amount of adulteration is practiced with snuff, and this in some instances of a most dangerous kind. Out of forty-three samples of snuff examined by Dr. Hassell, the majority were adulterated considerably. Chromate of lead, oxide of lead, and bichromate of potash, all highly poisonous, were detected. Mr. Phillips also stated to the committee of adulteration, that he had found in different samples common peat, such as is obtained from the bogs of Ireland, starch, ground wood of various kinds, especially fustic, extract of logwood, chromate of lead, bichromate of potash, and various ochreous earths. Samples of spurious snuff, it is presumed for the purpose of mixing, were found to be composed of sumach, umber, Spanish brown, and salt; another kind was made up of ground peat, yellow ochre, lime, and sand, all of these being more or less scented.

The numerous varieties of snuff owe their character principally to the peculiarity of scent and the method of preparation. The perfumes used are either the essential oil of bergamot or otto or roses, and in some cases powdered orris root or Tonquin beans. The powdered leaves of the sweet-scented woodruff and the fragrant melilot have been alluded to as used for the same purpose, also the dried leaves of some species of orchis (*Orchis fusca*, etc.)

As a substitute for snuff, either in preference, or in cases where tobacco snuff could not be readily obtained, different vegetable productions have come into use. In India the powdered rusty leaves of a species of rhododendron (*R. campanulatum*), and in the United States the brown dust found adhering to the petioles of several species of kalmia and rhododendron, all of which possess narcotic properties, are used for this purpose. The powdered leaves of asarabacca have been named as the base of some kind of cephalic snuff. "Grimstone's eye snuff" has long enjoyed a certain amount of popularity, although it does not contain a particle of tobacco, but is composed mainly of such harmless ingredients as powdered orris root, savory, rosemary, and lavendar.

But to return to the subject of deleterious adulteration, we find in Dr. Hassell's *Adulterations Detected in Food and Medicine* several pages occupied with this really important subject. First comes the narration of a case of slow poisoning, on the authority of Professor Erichsen, by means of snuff containing as an adulteration 1.2 per cent. of oxide of lead. Then follows the case of Mr. Fosbroke, of injuries sustained from snuff containing lead. These are followed by other instances showing that all the combinations of lead tested, exhibited dangerous and disastrous symptoms, if indulged in, when mingled with snuff, as too often, unfortunately, is the case, as an adulteration, or, as before shown, liable as a result of packing the snuff in lead, or keeping in boxes lined with lead.

ADVICE GRATIS—Give up taking snuff; or, if you should propose slight objections to this course, then purchase leaf tobacco, and manufacture your own snuff, and having done so, keep it in a gold snuff-box, or if you have weighty reasons for preferring silver, there is no objection to that metal, or even the homely horn of the Franciscan of Calais.

Our forefathers thought of the box, as well as of the snuff, and sometimes paid for their thought. In the early part of the eighteenth century, fashionable snuff-boxes had reached the highest point of luxury and variety. *The Tatler* of March 7, 1710, notices several gold snuff boxes which "came out last term," but that "a new edition would be put out on Saturday next, which would be the only one in fashion until after Easter. The gentleman," continues the notice, "that gave £50 for the box set with diamonds, may show it till Sunday, provided he goes to church, but not after that time, there being one to be published on Monday that will cost fourscore guineas." These costly articles, so happily satirized by Steele, are represented as the productions of a fashionable toyman, named Charles Mather, popularly known under the name of "Bubble Boy."

Nor must we forget the amber snuff-box of which Sir Plume, in the *Rape of the Lock*, was so justly vain; in 1711 he "spoke, and rapped the box." In 1733, Dodsley mentions boxes made of shell, mounted in gold and silver. Latterly we have made the acquaintance of several shell snuff-boxes; some of these were made of the tiger cowry, mounted in silver; of a small species of Turbo, cleaned and polished, and of harp shells, either mounted in silver or in baser metal. In different parts of the globe, tastes differ as to the materials of which snuff-boxes should be composed. A gentleman sent a piece of cannel coal from England to China, to be there carved by the ingenious Chinese into a snuff-box; this task was accomplished, and the box was shown in the Exhibition of 1851; also, in the Turkish department, a snuff-box of bituminous shale. Perhaps in the new Exhibition of 1862, there may be found a similar article, carved out of Gravesend flint, by natives of the Orange River Territory; or one of Suffolk coprolite, executed by rebellious sepoy women imprisoned in the hulks at Portsmouth.

In India, snuff-boxes are made of polished cocoa-nut shell, or of the seeds of *Entada gigalobium*, or *pursœtha*; or in Nepal, of a small kind of calabash or gourd, apparently resembling those used for the same purpose, at the distance of 5,000 miles, in the South of Africa; excepting, that in some instances, the gourds of Nepal and of Scinde, are ornamented with mounting of gold or silver, a luxury in which the African does not indulge. In the same part of Africa, among the Zoolu Kaffirs, other kinds of snuff-boxes, of smaller size, are in common use. These are made of the seeds of a species of Zamia, ornamented with strings of small beads, and are worn suspended as earrings, from the ears of the natives.

In China, flasks are used, the form and size of a smelling bottle; these are of different kinds of material, some being cut out of rock crystal, and others made of porcelain and similar plastic substances. Snuff-takers are less numerous in China than smokers of tobacco; in powder, or as the Chinese say, "smoke for the nose," is little used, except by the Mantchoo Tartars and Mongols, and Mandarins and lettered classes. The Tartars are real amateurs, and snuff is with them an object of the most important consideration. For the Chinese aristocracy, on the contrary, it is a mere luxury—a habit that they try to acquire—a whim. The custom of taking snuff was introduced into China by the old missionaries who resided at the Court. They used to get the snuff from

Europe for themselves, and some of the Mandarins tried it, and found it good. By degrees the custom spread; people who wished to appear fashionable, liked to be taking this "smoke for the nose"; and Pekin is still *par excellence*, the locality of snuff-takers. The first dealers in it made immense fortunes. The French tobacco was the most esteemed; and as it happened at this time, that it had for a stamp the ancient emblem of the three *fleur de lis*, the mark has never been forgotten, and the three *fleur de lis*, are still in Pekin, the only sign of a dealer in tobacco. The Chinese have now, for a long time, manufactured their own snuff, but they do not subject it to any fermentation, and it is not worth much. They merely pulverize the leaves, sift the powder till it is as fine as flour, and afterwards perfume it with flowers and essences. A curious method of snuffing, requiring neither box nor flask, is noticed in the *Voyages and Researches of the* Adventure *and* Beagle. At Otaheite, a substance, not unlike powdered rhubarb in appearance, but of a very pleasant fragrance, is rubbed on a piece of shark's skin stretched on wood; and an old man, who had one of these snuff sticks in his possession, valued it so highly, that he could not be induced to part with it.

Boxes of very rude construction are made in France and Germany from birch bark, and sold in the streets of Paris and other continental cities, for about one halfpenny each. These have lately been seen in the shops of London tobacconists under the name of "German boxes," at about three times the above price. They are used abroad either for tobacco or snuff. Boxes are also made of horn, either black buffalo or transparent pressed horn—the latter at a much cheaper rate than the former. St. Helena contributed to the Great Exhibition snuff-boxes made from the willow under which the remains of Napoleon reposed, until their removal to France, and also from a willow planted by him at Longwood. Van Dieman's Land contributed a box made from the tooth of the Sperm whale, as well as boxes from several native woods.

The Scotch snuff-boxes are justly celebrated for the perfection of their hinge, and close fitting cover. They were originally made at Lawrence-kirk, but the manufacture has now spread to various parts of Scotland. The wood employed principally in the manufacture of these boxes is the sycamore (or plane of the Scotch). Mr. W. Chambers states, "that from a rough block of this wood, worth twenty-five shillings, snuff-boxes may be made to the value of three thousand pounds."

The *modus operandi* in making these boxes is described as follows:

The box is made from a solid block of wood; the first operation consists in making a number of circular excavations in close contiguity to
each other, by means of a centre-bit, or a drill running in a lathe; the
interior is then squared out by means of gouges and chisels, and is
afterwards smoothed with files and glass-paper. The celebrated hinge
is formed partly out of the substance of the box, and partly out of that
of the lid, the greatest attention being paid in its construction to the
accurate fitting of the various parts one into the other. The box is
lined in the inside with stout tin-foil, and is painted on the outside
with several coats of colour, each of which is rubbed down smooth
with glass-paper before the succeeding coat is applied. It is then ready
to receive the various styles of ornament, which, in some cases, are
produced by the hand of the artist, and in others by mechanical means.
The most usual decoration consists of the tartan patterns, the component lines of which are drawn separately, by pens fixed in a ruling
machine, on to the box itself, if bounded by planes or slightly curved
surfaces; although such lines were also formerly drawn by means of a
rose engine on circular boxes, it is now found a more convenient practice to rule the lines on paper, and then to attach the paper to the
boxes. Another style of ornamentation, known as the Scoto-Russian,
is of more recent introduction, and imitates, in a remote degree, the
beautiful enamelled silver snuff-boxes for which Russia has long been
famous. In these, the outside of the box is first covered with stout tin-
foil, then completely painted all over the surface, and afterwards placed
in the ruling machine, which traces upon it an intricate pattern of
curved and straight lines, by means of a sharp flat tool. This instrument penetrates completely through the paint, but only scrapes the
tin-foil, which is left very bright, and resembles inlaid silver. Several
coats of copal varnish, each of which is successively polished down,
are then applied to complete the snuff-box.

Box-wood, box-root, king-wood, ebony, and all kinds of hard wood;
tin, brass, pewter, lead, silver, and all sorts of metals, are used for snuff-
boxes, some of these cheap and rudely fashioned, others elaborate
and expensive; some lined with tortoise-shell or horn, others with tin
or lead-foil; and invention has been taxed to produce all kinds of
ornamentation.

The practice of using snuff is said to have come into England after
the Restoration, and to have been brought from France; but it is well

known that the habit of mere snuff-taking did not originate with the introduction of tobacco, since there are recipes for making snuff from herbs in the oldest medicinal works extant. The use of tobacco snuff has been referred to the age of Catherine de Medici, and it was recommended to her son, Charles IX, for his chronic headaches. Snuff-taking was formerly characteristic of the medical profession; and the gold-headed cane and gold snuff-box came to be the peculiar emblems of those who were learned in the healing art.

There are almost an endless variety of snuffs, as of noses, the purest kind being the "Scotch," made either entirely from the stalks removed from the leaf in the course of its preparation for the cigar, or of the stalks with a small quantity of leaf. The "Welsh" and "Lundyfoot" are affirmed to owe their qualities chiefly, if not altogether, to the circumstance of their being dried almost to scorching; hence they have received the appellation of "high-dried" snuffs. The "Rappees" and other dark snuffs are manufactured from the darker and ranker leaves. Scenting, which the dark snuffs undergo, also furnish names and procure customers for numerous varieties. There is a story current, that the celebrated "Lundyfoot" had its origin in an accident, one version affirming that the man who was attending to the batches got drunk, neglected his duty, and made his master's fortune; another, that an accidental fire did that for the firm which in the other case it is affirmed that an extra glass of grog accomplished. There is nothing surprising in this, and either narrative may be true; most inventions of this kind, like the claying of sugar, had their origin in accidents. A certain quantity of snuff, in the preparation, gets overdone in some of the steps of the process, at some time or other, and the firm resolves, perhaps, as it is not altogether useless, to try and realize something for it. The peculiarity just tickles certain noses, and for the future they wish for none but *spoilt* snuff; that which was at first spoilt accidentally, is now spoilt for the purpose, to supply the demands of the market at even a higher rate than ordinary, and the name of Lundyfoot becomes immortalized amongst old ladies through all succeeding generations. What other experiments and other accidents of over-salting or over-liming may have done, has not transpired; and who may be the next so to turn circumstances to account, that what would ordinarily be considered a misfortune, shall be turned to good fortune, time alone will reveal.

John Hardham was Garrick's under-treasurer, and kept a snuff-shop in Fleet Street, at the sign of the Red Lion, where he contrived to get into high vogue, a particular *poudre de tabac*, still known as Hardham's 37. Stevens, while daily visiting Johnson in Bolt Court, on the subject of their joint editorship of Shakespeare, never failed to replenish his box at the shop of a man who was for years the butt of his witticisms. Hardham died a bachelor, September 20, 1772, and bequeathed £6000— the savings of a busy life—for the benefit of the poor of his native city, Chester.

As a pinch of snuff ends in a sneeze, so sniffing ends in sneezing, and with a hearty sneeze we bring our pinch of snuff to a sudden ending. What comfort and consolation there is sometimes in a hearty sneeze, no one knows better than him who has just made two or three attempts, and ingloriously failed. With half closed eyes, and open mouth, and bated breath—once—twice—thrice—no! it will not be beguiled—psh-h-h-h-haw! "God bless you!"

"The year 750," says a writer in the *Gentleman's Magazine*, "is commonly reckoned the era of the custom of saying God bless you to one who happens to sneeze." It is said that, in the time of the pontificate of St. Gregory the Great, the air was filled with such a deleterious influence, that they who sneezed immediately expired. On this the devout pontiff appointed a form of prayer, and a wish to be said to persons sneezing for averting them from the fatal effects of this malignancy. A fable contrived against all the rules of probability, it being certain that this custom has from time immemorial, subsisted in all parts of the known world. According to mythology, the first sign of life Prometheus's artificial man gave, was by sternutation. This supposed creator is said to have stolen a portion of the solar rays, and filling a phial with them, sealed it up hermetically. He instantly flew back to his favourite automaton, and opening the phial, held it close to the statue, the rays still retaining all their activity, insinuated themselves through the pores, and set the factitious man a sneezing. Prometheus transported with success, offered up a prayer with wishes for the preservation of so singular a being. The automaton observed him, remembering his ejaculations, was careful, on like occasions to offer these wishes in behalf of his descendants, who perpetuated it from father to son in all their colonies. The Rabbis, also, fix a very ancient date to the custom. Pliny says, that to sneeze to the right was deemed fortunate; to the left, and near a place of burial, the reverse. Tiberius, otherwise a sour man, would perform this right of blessing most punctually to others, and expect the same from others to himself. Aristotle has a problem, "Why sneezing from noon to midnight was good, but from night to noon unlucky." St. Austin tells us that the ancients were accustomed to go to bed again, if they sneezed while they put on their shoe.

When Themistocles sacrificed in his galley before the battle of Xerxes, one of the assistants upon the right hand sneezed, Euphrantides the soothsayer, presaged the victory of the Greeks, and the overthrow of the Persians.

When the Greeks were consulting concerning their retreat in the time of Cyrus the Younger, it chanced that one of them sneezed, at the noise whereof, the rest of the soldiers called upon Jupiter Soter.

Brand tells us, that when the king of Mesopotamia sneezes, acclamations are made in all parts of his dominions. The Siamese wish long life to persons sneezing. And the Persians look upon sneezing as a happy omen, especially when repeated often.

A writer lately gives us the following "Philosophy of a sneeze" for which he alone is responsible.

> The nose receives three sets of nerves—the nerves of *smell*, those of *feeling*, and those of *motion*. The former communicate to the brain, the odorous properties of substances with which they may come in contact, in a diffused or concentrated state; the second, communicate the impression of touch; the third, move the muscles of the nose; but the power of these muscles is very limited. When a sneeze occurs, all these faculties are excited to a high degree. A grain of snuff excites the olfactory nerves, which despatch to the brain the intelligence that 'snuff has attacked the nostril.' The brain instantly sends a mandate through the motor nerves to the muscles, saying 'cast it out!' and the result is unmistakable. So offensive is the enemy besieging the nostril held to be, that the nose is not left to its own defence. It were too feeble to accomplish this. An allied army of muscles join in the rescue—nearly one-half the body arouses against the intruder—from the muscles of the lips to those of the abdomen, all unite in the effort for the expulsion of the grain of snuff.

CHAPTER VII

QUID PRQ QUQ

A third party sprang up, headed by the descendants of Robert
Chewit, the companion of the great Hudson. These discarded
pipes altogether, and took to chewing tobacco; hence, they
were called Quids.

Knickerbocker's New York

Any one who will take the trouble to
read through the *Curiosities of Food*, will soon become convinced, from
the examples which Mr. P. L. Simmonds has collected so assiduously
from all parts of the world, that there is no accounting for tastes. What
extraordinary things men will admit between their teeth to gratify
their appetites, is almost enough to set one's own teeth on edge. Tobacco is certainly not more nauseous or revolting, than to us would be
many of the delicacies dished up for dinner by some of the bipedal
race. "Some Europeans," observes the author,

chew tobacco, the Hindoo takes to betel nut and lime, while the
Patagonian finds contentment in a bit of guano, and the Styrians
grow fat and ruddy on arsenic. English children delight in sweetmeats and sugar-candy, while those in Africa prefer rock salt. A
Frenchman likes frogs and snails, and we eat eels, oysters, and whelks.
To the Esquimaux, train oil is your only delicacy. The Russian luxuriates upon his hide and tallow; the Chinese upon rats, puppy dogs,
and shark fins; the Kaffir upon elephant's foot and trunk or lion
steaks; while the Pacific islander places cold missionary above ev-

ery other edible. Why then should we be surprised at men's feeding upon rattle snakes and monkeys, and pronouncing them capital eating?*

Nothing is more extraordinary than the habit of dirt-eating and chewing of lime, either by themselves or in combination with other substances. But more of this anon. Tobacco, as a masticatory, might equally cause surprise did it not daily occur at our doors. The quantity used in this form will not bear comparison with that consumed in smoke, but even this is considerable. In America, the custom is carried to a very unpleasant extent, and were it the only form in which the plant could be indulged, there is good ground for presuming that it would fall very far short of the popularity which it has attained.

Somebody, with a strong antipathy to pig-tail and fine cut, has entered into certain investigations and calculations in the *Philadelphia Journal*, which has resulted in this wise. If a tobacco chewer chews for fifty years, and uses each day of that period two inches of solid plug, he will consume nearly one mile and a quarter in length of solid tobacco, half an inch thick and two inches broad, costing 2,094 dollars, or about £500. Plug ugly, sure enough! By the same process of reasoning, this statist calculates, that if a man ejects one pint of saliva per day for fifty years (a feat, one would presume, it would require a Yankee to accomplish), the total would swell into nearly 2,300 gallons, quite a respectable lake, and almost enough to float the *Great Eastern* in! Truly, Brother Jonathan, there are more things in heaven and earth than are dreamt of in our philosophy.

Another calculation shows, that if all the tobacco which the British people have consumed during the last three years were worked up into pig-tail half an inch thick, it would form a line 99,470 miles long; or enough to go nearly four times round the world;[†] or if the tobacco consumed by the same people in the same period were to be placed in

---

Curiosities of Food, by P. L. Simmonds. Bentley, 1859.
[†]Tobacco entered for home consumption—

1856	1857	1858
32,579,166 lbs.	32,851,365 lbs.	34,110,850 lbs.
	Total 99,541,381 lbs.— or	44,438 tons.

one scale, and St. Paul's Cathedral and Westminster Abbey in the other, the ecclesiastical buildings would kick the beam.

"Oh, the nasty creatures!" some lady exclaims. "Who could suppose that they would do such a thing, and to such an extent too, as to burn and chew and smoke in three years enough tobacco to reach round the world four times!" It is astonishing, my dear Mrs. Partington, we must confess; but let us compare therewith the tea consumption* for the same period, and we shall find that during the past three years, we have consumed about 205,500,000 of pounds of tea, which, if done up in packages containing one quarter of a pound each—such packages being $4^1/_2$ inches in length and $2^1/_2$ inches in diameter—these placed end to end, would reach 59,428 miles; or, upon the same principles as those adopted for the pig-tail, would girdle the earth twice with a belt of tea $2^1/_2$ inches in diameter, or twenty-five times that of the aforesaid pig-tail. Enough to make rivers of tea strong enough for any old lady in the kingdom to enjoy, and deep enough for all the old ladies in the kingdom to bathe in.

All this, we are free to confess, does not make the habit of quidding either more justifiable or respectable, although indulged in by some of the members of the gentler sex. In Paraguay, for instance, an American traveller informs us that everybody smokes, and nearly every woman and girl more than thirteen years old chews tobacco. A magnificent Hebe, arrayed in satin and flashing in diamonds, puts you back with one delicate hand, while with the fair taper fingers of the other she takes the tobacco out of her mouth previous to your saluting her. An over delicate foreigner turns away with a shudder of loathing under such circumstances, and gets the epithet of "the savage" applied to him by the offended beauty for his sensitive squeamishness. However, one soon gets used to these things in Paraguay, where one is, per force of custom, obliged to kiss every lady one is introduced to, and one half of those you meet are really tempting enough to render you reckless of consequences.

Suppose not that Paraguay is a solitary instance in which ladies have a predilection for this masticatory. In Siberia, which is far enough

---

*Tea entered for home consumption in—

1856	1857	1858
63,295,643 lbs.	69,159,640 lbs.	73,217,483 lbs.

geographically to prevent any collusion, or the influence of example to exert its power, Captain Cochrane says that the Tchuktchi eat, chew, smoke, and snuff at the same time. He saw amongst them, boys and girls of nine or ten years of age who put a large leaf of tobacco into their mouths without permitting any saliva to escape, nor would they put aside the tobacco should meat be offered to them, but continued consuming both of them together.

The Mintira women and other races of the great Indian Archipelago are addicted to chewing tobacco. Amongst the Nubians, the custom is more common than smoking. Of the South American tribes, the Sercucumas of the Erevato, and the Caura neighbours of the whitish Taparitos, swallow tobacco chopped small, and impregnated with some other stimulant juices.

In Africa, the habit is not at all an uncommon one. The Turks and Arabs of Egypt are great smokers, but not so with the other tribes. The Mongrabins, scarcely know the use of a pipe, or the method of manufacturing a cigar, yet tobacco is well known, and chewing is the order of the day. With them each piece of tobacco is mixed with a portion of natron. Master and servant, rich and poor, all carry about them a pouch of tobacco, with pieces of natron in it. These people do not carry the quid in their cheek, as do the Europeans who indulge in the habit, but in front, between the teeth and the upper lip.

The blacks of Gesira have another method of enjoying this luxury. They make a cold infusion of tobacco, and dissolve the natron in it. This mixture is called "bucca." The natives take a mouthful of it from the bucca cup, which they keep rinsing and working about in their mouths for a quarter of an hour before they eject it. So much do they delight in it, that it is considered the highest treat a man can offer to his dearest friends, to invite them to sip the bucca with them. Bucca parties are given, as in some localities tea parties are honoured. All sit in solemn silence as the cup goes round, each taking a mouthful, and nothing is heard save the gurgling and working inside the closed mouths. On such occasions the most important questions receive no reply, for to open the mouth and answer would be to lose the cherished bucca.

In Iceland, tobacco is chewed and snuffed as assiduously as it is smoked in other countries; and in the northern states of Europe, or some of them, the powdered leaf, which, with most people is deemed a preparation for the nose, is placed, a pinch at a time, upon the tongue.

Of Joubert's statement we scarce know what opinion to hold. He says, "When a stranger arrives in Greenland, he is immediately surrounded by a crowd of the natives, who ask the favour of sucking the empy-reumatic oil in the reservoir of his pipe. And it is stated that the Greenlanders smoke only for the pleasure of drinking that detestable juice which is so disgusting to European smokers." The Finlander de-lights in chewing. He will remove his quid from time to time, and stick it behind his ear, and then chew it again. This reminds us of a circum-stance narrated by a friend, which occurred when he was a boy. His master was a chewer. After a quid had been masticated for some time, it was removed from his mouth, and thrown against the wall, where it remained sticking; the apprentice was then called to write beside it the date at which it was flung there, so that it might be taken down in its proper turn, after being thoroughly dried, to be chewed over again.

> And then he tried to sing All's well,
> But could not though he tried;
> His head was turned, and so he chewed
> His pig-tail till he died.

Of all tobacco chewers, none can compete with the Yankee—not even our own Jack Tars. They are the very perfection of masticators, and of spitters, also, if the narratives of travellers in general, and of Dickens in particular, are to be relied on.

> As Washington may be called the head-quarters of tobacco-tinc-tured salvia, the time is come when I must confess, without any disguise, that the prevalence of these two odious practices of chew-ing and expectorating began, about this time, to be anything but agreeable, and soon became the most offensive and sickening. In all the public places of America, this filthy custom is recognized. In the courts of law, the judge has his spittoon, the crier his, the wit-ness his, and the prisoner his, while the jurymen and spectators are provided for, as so many men who, in the course of nature, must desire to spit incessantly. In the hospitals, the students of medicine are requested by notices upon the wall, to eject their tobacco juice into the boxes provided for that purpose, and not to discolour the stairs. In public buildings visitors are implored, through the same agency, to squirt the essence of their "quids" or "plugs," as I have

heard them called by gentlemen learned in his kind of sweetmeat, into the national spittoons, and not about the bases of the marble columns. But in some parts this custom is inseparably mixed up with every meal and morning call, and with all the transactions of social life. The stranger who follows in the track I took myself, will find it in its full bloom and glory at Washington; and let him not persuade himself (as I once did to my shame) that previous tourists have exaggerated its extent. The thing itself is an exaggeration of nastiness which cannot be outdone.

On board the steamboat there were two young gentlemen, with shirt collars reversed, as usual, and armed with very big walking sticks, who planted two seats in the middle of the deck, at a distance of some four paces apart, took out their tobacco boxes, and sat down opposite each other to chew. In less than a quarter of an hour's time, these hopeful youths had shed about them on the clean boards, a copious shower of yellow rain, clearing by that means a kind of magic circle, within those limits no intruders dared to come, and which they never failed to refresh and refresh before a spot was dry. This being before breakfast, rather disposed me, I confess, to nausea; but looking attentively at one of the expectorators, I plainly saw that he was young at chewing and felt inwardly uneasy himself. A glow of delight came over me at this discovery, and as I marked his face turn paler and paler, and saw the ball of tobacco in his left cheek quiver with his suppressed agony, while yet he spat and chewed, and spat again, in emulation of his older friend, I could have fallen on his neck and implored him to go on for hours.

The senate is a dignified and decorous body, and its proceedings are conducted with much gravity and order. Both houses are handsomely carpetted; but the state to which these carpets are reduced by the universal disregard of the spittoon, with which every honorable member is accommodated, and the extraordinary improvements on the pattern which are squirted and dabbled upon it in every direction, do not admit of being described. I will merely observe, that I strongly recommend all strangers not to look at the floor; and if they happen to drop anything, though it be their purse, not to pick it up with an ungloved hand on any account. It is somewhat remarkable to discover, that this appearance is caused by the quantity of tobacco they contrive to stow within the hollow of the cheek. It is strange enough, too, to see an honorable gentleman leaning back in his tilted chair, with his legs on the desk before him, shaping a convenient "plug" with his penknife, and when it is quite ready for use, shooting the old one from his mouth as from a pop-gun, and clapping the new one in its place. I was surprised to observe, that even steady old chewers of great experience are not always good marksmen, which has rather inclined me to doubt that general proficiency with the rifle of which we have heard so much in England. Several gentlemen called upon me, who, in the course of conversation, frequently missed the spittoon at five paces; and one (but he was certainly short-sighted) mistook the closed sash for the open window at three. On another occasion when I dined out, and was sitting with two ladies and some gentlemen round a fire before dinner, one of the company fell short of the fireplace six distinct times. I am disposed to think, however, that this was occasioned by his not aiming at that object, as there was a white marble hearth before the fender, which was more convenient, and may have suited his purpose better.

At the Cape of Good Hope grows a plant, allied to the iceplant of our greenhouses, and which is a native of the Karroo,* which appears to possess narcotic properties. The Hottentots know it under the name of Kou, or *Kauw-goed*. They gather and beat together the whole plant, roots, stem, and leaves, then twist it up like pig-tail tobacco; after which they let the mass ferment, and keep it by them for chewing, especially when they are thirsty. If it be chewed immediately after fermentation, it is narcotic and intoxicating. It is called canna-root by the colonists.

In Lapland, Angelica-root (*Archangelica officinalis*, Linn.) is dried and masticated in the same way, and answers the same purpose as tobacco. It is warm and stimulating, and not narcotic, nor does it leave those unpleasant and unsightly evidences of its use which may be observed about the mouth of the true votary of the quid.

The areca nut and the betel-pepper, which, in the Malayan Peninsula and other parts of the East, are used as a masticatory, will receive special notice hereafter.

Mesembryanthemum tortuosum, Linn.

Lightfoot says that the Scotch are very fond of "dulse," but they prefer it dried and rolled up, when they chew it like tobacco, for the pleasure arising from the habit. This is the only reference to the custom that we have met with, and requires further confirmation.

The Duke of Marlborough has the credit of being the first distinguished man who made the chewing of tobacco famous; who was the last is not so readily declared, since distinguished men generally do not distinguish themselves much in this department of the "fine arts." It is related of a monkey, that while on the voyage home from some tropical clime in which he had been made a prisoner, he noticed a sailor who was in the habit of going to his trunk and taking out a quid, roll it up, and place it in his mouth. Finding, one day, that the course was clear, and the box unfastened, Jocko helped himself to a very respectable twist, which he put into his mouth, and scampered therewith upon the deck. He soon commenced chewing and spitting, and, unsuccessful in the experiment, the quid, which was not found to be so pleasant as anticipated, was thrown away. The poor animal soon became dreadfully sick, held its stomach, and moaned piteously, but ultimately recovered. He learnt a lesson, however, the impression of which never passed away; for ever after he shunned the box, and the sight or smell of tobacco sent him scampering into the shrouds.

CHAPTER VIII
A RACE OF PRETENDERS

*I grant your worship that he is a knave, sir; but yet, Heaven
forbid, sir, but a knave should have some countenance at his
friends request. An honest man, sir, is able to speak for
himself, when a knave is not.*

King Henry IV, part 2

It is the misfortune of kingdoms to be
subject to rebellions, and of monarchs to behold the advent of pretenders, as it is the fate of gold to be imitated in baser metals, and
bank notes to be forged. A rule is supposed to be strengthened by an
exception, and tried gold to shine in greater splendour beside its
counterfeit—

Than that which hath no foil to set it off.

So, tobacco, in the midst of all its success and prosperity, has been
envied and imitated by duller pretenders to the virtue it boasts, from
among the meaner denizens of the vegetable world. Of course these
pretenders have been unsuccessful; for had they been successful, they
had no longer been branded with the baser name, but had risen to the
rank of benefactors and patriots. Such is the custom of the world.

The following are the substances which are stated to be used for
the adulteration of tobacco, principally in the form of "cut" and "roll."
Dr. Hassell divides them—

First, into vegetable substances, as the leaves of the dock,
rhubarb, coltsfoot, cabbage, potato, chicory, endive, elm,

and oak; malt cummings, that is the roots of germinating malt; peat, which consists chiefly of decayed moss;
seaweed, roasted chicory root, wheat, oatmeal, bran, catechu or terra japonica, oakum, and logwood dye.

Secondly, into saccharine substances, as cane-sugar, treacle,
honey, liquorice, and beetroot dregs.

Thirdly, into salts and earths, as nitre, common salt, sal am-
moniac, or hydrochlorate of ammonia, nitrate of ammonia, carbonate of ammonia, the alkalies, as potash, soda,
and lime; sulphate of magnesia, sulphate of soda or
glauber salts, yellow ochre, umber, fuller's earth, Vene-
tian red, sand, and sulphate of iron.

And the experience of the excise, as may be gathered from the
evidence of Mr. Phillips before the committee of adulteration, har-
monizes with the above list. "With regard to tobacco," he says, "we
have found in *cut* tobacco, sugar, liquorice, gum catechu, saltpetre,
and various nitrates; yellow ochre, Epsom salts, glauber salts, green
copperas, red sandstone, wheat, oatmeal, malt cummings, chicory,
and the following leaves—coltsfoot, rhubarb, chicory, endive, oak,
elm; and in *fancy* tobacco, I once found lavender, and a wort called
mugwort. It is a fragrant herb, suggestive rather of the nutmeg. In
roll tobacco we have found rhubarb leaves, endive and dock leaves,
sugar, liquorice, and a dye made of logwood and sulphate of iron."

Let consumers of tobacco console themselves, however, in the face
of this formidable list, by the assurance of the eminent experimenter
on articles of food, etc., before named, that "not one of the forty samples
of manufactured cut tobacco which he examined was adulterated with
any foreign leaf, or with any insoluble or organic extraneous substance
of any description other than with sugar, or some other saccharine
matter, which was present in several instances."

Leaving adulterations to take care of themselves, we find that
an article, of very ancient use, is still occasionally smoked instead
of the Virginian weed. The plant referred to is *coltsfoot* (*Tassilago
farfar*, Linn.), a very common weed on chalky and gravelly soils.
Pliny refers to it, and directs that the foliage should be burned,
and the smoke arising from it drawn into the mouth through a
reed and swallowed. These leaves have long been smoked for chest

complaints, and are said to form the chief ingredient in British herb tobacco.

The leaves of milfoil or yarrow (*Achillœa millefolium*), another plant equally common with the last, have been recommended to smokers in lieu of tobacco, and occasionally used for that purpose. Added to beer, they render it heady or more intoxicating.

Leaves of rhubarb are occasionally smoked by those who are too poor to furnish themselves with a regular supply of tobacco, and those who have used them state, that, although devoid of strength, they are not a bad substitute when tobacco is not to be obtained. For the same purpose they are collected and used in Thibet, and on the slopes of the Himalayas.

The leaves of a plant common in marshes and boggy soils in Europe and North America, called bogbean (*Menyanthes trifoliata*, Linn.) are used in the north of Europe when hops are scarce, to give a bitter flavour to beer, and have been recommended and adopted as a tobacco substitute.

An agricultural labourer near Blois, pretends that the leaves of the beet make an excellent tobacco.

Undescribed plants called Akil and Trouna, are used by the Arabs of Algeria to render their tobacco milder.

In some parts of Europe, the leaves of the common garden sage has served the same purpose; whilst in some parts of Switzerland, the leaves of mountain tobacco (*Arnica montana*, Linn.) are collected for use as tobacco, or dried and powdered to be used as snuff. This is no doubt a virulent plant, and has the reputation of being a powerful acrid narcotic.

The tobacco substitutes in North America are more numerous than we should have expected to have found in the native land of the true tobacco. A decoction of the holly-leaves (*Ilex vomitoria*, Linn.) are drunk by the native Creek Indians, under the name of "black drink," at the opening of their councils, on account of its peculiar properties. This shrub is also called cossena by the Indians, and the leaves are used for smoking as a substitute for tobacco. "Often," says one of the early settlers, "I have smoked a pipe of cossena with their majesties Toma Chaci and Senoaki his queen, at their mud-palace, about three miles from Savanacke."

The Virginian or Stag's Horn Sumach,* which is met with almost over the whole of the United States, supplies leaves which are dried and used by some of the native tribes as tobacco.

The Indians of the Mississippi and Missouri use the leaves of another sumach (Rhus copallina) and Indian tobacco (Lobelia inflata, Linn.) is supposed to be indebted for its name to the fact that it was one of the plants smoked by the Indians instead of the genuine "weed." Under the name of "tom-beki," the leaf of a species of Lobelia is smoked in parts of Asia. It is smoked in a narghilè, and is exceedingly narcotic, so much so, that it is usually steeped in water to weaken it before being used; and it is always smoked whilst damp.

Not many years since, a patent was taken out at Washington for fabricating tobacco from maize-husks, steeped in a solution of cayenne. It was stated to be equal in flavour to true tobacco, and without any of the deleterious properties which have been attributed to that plant.

The Miliceti Indians, New Brunswick, scrape the bark from the twigs of the birch, and when dry, mix it with their tobacco for smoking. They are very partial to the admixture, the odour of which, it is affirmed, is much more agreeable than that of pure tobacco.

Mr. Mölhausen smoked willow-leaves among the Rocky Mountains; and the use of these leaves for the same purpose is mentioned in Hiawatha.

The bearberry (Arctostaphylus uva ursi) common in many parts of North America, is found in the valley of the Oregon, where the leaves are collected by the Chenook Indians, who mix them with their tobacco. The Crees also use them for the same purpose, and with them it is called Tchakashè-pukh. The Chepewyans, who name it Kleh, and the Eskimos north of Churchill (by whom it is termed Attung-ā-wi-at) turn it to a like account. From the custom of the Hudson's Bay Company's officers carrying it in bags for the same use, the voyagers gave it the appellation of Sac-a-commis.

Latterly a writer in a West Indian paper, called attention to a novel application of the berries of the pimento (Eugenia pimento), known

---

*Rhus typhina.

commercially by that name or as Allspice. "I have been," he says,

> a smoker for the past twenty years, and have consumed many pounds
> of honey-dew within that period; but it was only a short time ago
> that I discovered that Pimento forms by far a more agreeable article
> for smoking; and any person who knows nothing of the fragrance of
> a Pimento walk when in full bloom, may form some idea of it by a
> pipe charged and lighted with the dried berry, simply crushed in
> coarse bits. Every lady has a dislike to the smell of tobacco. While
> she may be driven by its fumes and smell from the drawing-room,
> the Pimento would, on the contrary, invite her presence. By way of
> experiment on the taste of other smokers, I may mention that I had
> the other day two men (great lovers of tobacco) employed in my
> garden. "Joseph," I said, "where is your pipe to-day?" "Out of to-
> bacco, massa," was his reply. "Well, here is some very costly; give
> me your opinion of it when you have tried it." To prevent decep-
> tion, I charged his pipe myself, and directed him to light it. He did
> so, and up ascended a graceful curl of smoke. Joseph was not a little
> pleased, and thanking me for this costly tobacco, said it was "first-
> rate," and desired I should inform him what per pound it could have
> cost. I told him it grew pretty near his hut, and on opening my
> pouch, and disclosing to him that this "first-rate tobacco" was noth-
> ing more than dried pimento, you may imagine his surprise. "A man
> is neber too old to larn," he exclaimed, and soon imparted the good
> news to his fellow-labourer.

With all due deference to the opinion of both Joseph and his master,
we have experimented on this wonderful pretender, and hold the opinion that it is unworthy of their joint encomiums. A friend who has
also tested it, thinks it, however, very pleasant, and a fair substitute. It
would appear, therefore, that there is something to be said on both
sides.

Cascarilla bark, the produce of the *Croton eleuteria* in the Bahamas,
was first used to mix with tobacco, on account of the pleasing odour
which it diffuses in burning. It is supposed also to possess narcotic properties, when used in this way. In South America, Humboldt states that
the leaves of *Polygonum hispida* are used as a tobacco substitute.

The African contributions to our list are also rather extensive,
especially from the neighbourhood of the Cape. The leaves of a cer-

tain plant (*Tarchonanthus camphoratus*, Linn.) possessing a camphorated odour, are chewed by the Mahometans, and smoked by the Hottentots and Bushmen instead of tobacco, and, like the *Dagga*, exhibit slight narcotic symptoms. This may be owing to the camphor which they contain. The common camphor, in quantities a little beyond a medium dose, will produce indistinctness of ideas, incoherence of language, an indescribable uneasiness, shedding of tears, a sensation of fear and dread; then the body feels lighter than usual—an idea exists that flying will not only be easy, but a source of pleasure.

The Wild Dagga (*Leonotis leonurus*, R. Br.) grows wild on the sandy Cape flats. It has a peculiar scent, and a nauseous taste, and seems to produce narcotic effects if incautiously used. The Hottentots are particularly fond of it, and smoke it as tobacco. In the eastern districts of the Cape, an allied species (*Leonotis ovata*) has a similar reputation, and is used for a like purpose.

In the Mauritius the leaves of the *Culen* (*Psoralea glandulosa*) are dried and smoked, while on the western coast of South America they are used in decoction as a beverage, instead of tea.

In Asia, tobacco substitutes have but one or two representatives. One of these has been already alluded to, another consists of the long leaves of a species of *Tupistra*, called "Purphiok," which are gathered in Sikkim, chopped up, and mixed with tobacco for the hookah. The leaves of the water-lily are dried, and used in China to mix with tobacco for smoking, to render it milder.

Cigars of stramonium, henbane, and belladonna, may be purchased at the same rate as those made of genuine tobacco, in chemists' and herbalists' shops—never having tried them, we have no experience of their flavour.

The majority of the substitutes for tobacco are, after all, very poor pretenders—capable, perhaps, of raising a smoke, but possessed of neither aromatic nor stimulating properties; and those which contain any active properties at all, are of a character so dangerous, as to make their extensive use extremely hazardous. In the former class, we may rank coltsfoot, sage, milfoil, rhubarb, and bogbean; and in the latter, stramonium, henbane, belladonna, arnica, and lobelia. Those who have been long accustomed to the use of tobacco, seldom, except in times of scarcity or deprivation of that plant, resort to the use of any other. This is the case at home. In the Cape Colony, the united testimony of travellers proves that the Kaffirs are ready to make *any* sacrifices for tobacco, and prefer it to any of their own indigenous substitutes.

When the tobacco has been found to be too strong, incipient smokers have been known to counteract its effects, and lessen its power, by mixing therewith the flowers of chamomile, which once enjoyed great reputation as a useful medicine. Others, in the absence of tobacco, have resorted to brown paper or tow, which, being smoked through an old or foul pipe, is said to carry with its smoke some of the tobacco flavour, and to be infinitely better than no smoke at all. Juveniles will sometimes, with a piece of cane, or a strip of clematis, imitate their elders, and, in imagination, enjoy the luxury of a Havannah cigar.

A curious anecdote of a Buckinghamshire parson occurs in *Lilly's History of His Life and Times*, to which we have before referred. "In this year, also, William Breedon, parson or vicar of Thornton in Bucks, was living, a profound divine, but absolutely the most polite parson for nativities in that age, strictly adhering to Ptolemy, which he well understood; he had a hand in composing Sir Christopher Heydon's *Defense of Judicial Astrology*, being at that time his chaplain; he was so given over to tobacco and drink, that when he had no tobacco (and I suppose too much drink) he would cut the bell-ropes and *smoke* them."

Having unmasked "the race of pretenders," and shown the titles upon which they seek to establish their claims, with Charles Lamb we now bid farewell to Tobacco.

> For I must (nor let it grieve thee,
> Friendliest of plants, that I must) leave thee;
> For thy sake, Tobacco, I
> Would do anything but die;
> And but seek to extend my days
> Long enough to sing thy praise.
> But as she, who once hath been
> A king's consort, is a queen
> Ever after, nor will bate
> Any title of her state,
> Though a widow, or divorced,
> So I, from thy converse forced,

The old name and style retain,
A right Katherine of Spain;
And a seat, too, 'mongst the joys
Of the blest Tobacco boys;
Where, though I, by sour physician,
Am debarred the full fruition
Of thy favours, I may catch
Some collateral sweets, and snatch
Sidelong odours, that give life,
Like glances from a neighbour's wife;
And still live in the by-places,
And the suburbs of thy graces;
And in thy borders take delight,
An unconquered Canaanite.

CHAPTER IX

"MASH ALLAH!"— THE GIFT

Farewell ye odours of earth that die,
Passing away like a lover's sigh;
My feast is now of the Tooba tree,*
Whose scent is the breath of eternity.

Moore's *Lalla Rookh*

That opium is the milky juice of the cap-
sules of a species of poppy, evaporated by exposure to light and air, is a
fact so well known, as scarce to require repetition. This species of
poppy contains two well marked varieties, the *black* and the *white*, a
circumstance noticed by Hippocrates long enough ago. The black
variety derives its name from the colour of its seeds. The original home
of the poppy is Asia and Egypt. But it is extensively cultivated for the
sake of its juice in British India, Persia, Egypt, and Asia Minor, and
might be cultivated, were it more remunerative, in England, France,
and Germany, where good samples of opium have been obtained experimentally. Dr. Royle states that the black variety is cultivated in
the Himalayas, but generally the white is preferred. The poppy is grown
in Europe for the sake of the capsules and seed; from the latter a mild
oil is extracted.

The cultivation of the poppy in British India is confined chiefly
to the large Gangetic tract, about six hundred miles in length, and

---

*"The tree Tooba that stands in Paradise, in the palace of Mahomet."—
Sale. "Tooba signifies beatitude or eternal happiness." —*D'Herbelot*.

two hundred miles in depth, extending from Goruckpore in the north to Hazareebaugh in the South; and from Dingepore in the East, to Agra on the West. This extent of country contains the two agencies of Behar and Benares, the former sending to the market about treble the quantity of the latter. In the Benares agency, there are about 21,500 cultivators, and the total number of under cultivators of the opium poppy 106,147.

After all the preliminaries of preparing the land, sewing, and cultivating the plant, all of which are much more interesting to the parties concerned than ourselves, if all goes well, the whole field of poppies presents a sheet of white bloom, which generally occurs about the month of February. When nearly ready to fall, the white petals are gathered, and made into circular cakes; these are preserved to form the outer coverings of the balls of opium. In a few days after the "leaves" of the flower are collected, the capsules or poppy heads are ready for operation. At from three to four o'clock in the afternoon, individuals go into the fields and scratch or cut the poppy heads with iron instruments called "nushturs." This instrument consists of three or four thin narrow strips of iron, about six inches in length, and about the thickness and width of a pen-knife at one end, but extending in width to nearly an inch at the opposite extremity, where it is deeply notched. These plates are bound together by means of thread, each plate being kept a little distance from its neighbour by means of thread passed between them. Thus completed, it has the appearance of a scarificactor with four parallel blades. This instrument, which has the angles sharpened, has one of its sets of points drawn down the poppy capsule from top to bottom, or rather upwards from the base to the summit, making three or four parallel incisions, corresponding to the number of blades in the poppy head. These only pass through the outer coating or pericarp. Each capsule is scarified from two to six times, according to its size, two or three days intervening between each operation. In Asia Minor, a different course is pursued. One horizontal incision is made nearly round the capsule, with a single blade. After the scarification of the capsules, the juice exudes and thickens on them during the night, which is collected early the next morning, by means of little iron instruments called "seetooahs," and which resemble small concave trowels. When sufficient is collected into the trowel, it is emptied into an earthen pot which the collector carries at his side.

When all the opium is collected which the plants will yield, the capsules are gathered and broken, and the seed preserved for the extraction of their oil. Of these seeds comfits are also made resembling carraway comfits, and, without doubt, great comforts they are to naked little squalling Hindoos whenever they can be obtained. After the extraction of the oil, the dry cake, called Khari, is either made into unleavened cakes for the very indigent, or cattle are fed upon them, or when necessity requires, it is converted into poultices after the manner of linseed meal.

In poor districts, where the people cannot afford the luxury of opium, the broken capsules are made into a decoction and drank instead, says Mr. Impey. This liquid is termed "post," from the Persian name of the capsule. There is also another use for the capsules. They are ground into fine powder, and sold under the name of "boosa," and sprinkled over the *buttees* of opium to prevent their adhesion. In the Benares agency, the stems and leaves, when perfectly dry, are collected and crushed into a course powder called "poppy trash" which is employed in packing the opium cakes.

One acre of well-cultivated ground will yield from 70 to 100 pounds of "chick" or inspissated juice, the price of which varies from six shillings to twelve shillings per pound; so that an acre will yield from twenty to sixty pounds worth of opium at one crop. Three pounds of chick will produce one pound of opium, from a third to a fifth of the weight being lost in evaporation.

When freshly collected, the mass of juice is of a pinkish colour. This is placed in shallow vessels to drain. A coffee-coloured liquid, called *pussewah*, is drained off, which is used to cement the poppy-leaves round the cakes of opium, under the name of *lewah*. After exposure to the air in the Benares agency, the opium is made up into balls. In Turkey it is the custom to beat up the juice with saliva. In Malwa it is immersed as collected in linseed oil. In Benares it is brought to the required consistence by exposure in the shade only.

Opium is prepared in different forms, in the various localities for market. Bengal opium is made into balls of about $3^1/2$ lbs. weight, and packed in chests, each containing forty balls. They are about the size of a child's head, coated externally with poppy petals, agglutinated with *lewah* to the thickness of about half an inch. Garden Patna opium is in square cakes, about three inches in diameter, and one inch thick, wrapped in thin plates of mica. Malwa opium is in round flattened cakes, of about ten ounces in weight, packed in "boosa," or in coarsely-powdered poppy-petals, or in some instances without any coating at all. Cutch opium is in small cakes, rather more than an inch in diameter, enclosed in fragments of leaves. Kandeish opium is imported in round flattened cakes, of about half a pound weight. Egyptian opium occurs in round flattened cakes, about three inches in diameter, covered with the vestiges of some leaf. This kind is very dry, but it is considered inferior in quality to the Turkish kinds. Persian opium is in the form of sticks, about six inches in length, and half an inch in diameter, enveloped in smooth shining paper, and tied with cotton. Smyrna opium occurs in regular rounded or flattened masses, of various sizes, rarely exceeding two pounds in weight, sometimes covered with the capsules of a species of dock. Constantinople opium is either in large irregular cakes, or small, regular, lenticular-formed cakes, covered with poppy-leaf, and from two to two and a half inches in diameter.

Formerly the balls of Bengal opium were covered with tobacco-leaves; but Mr. Flemming introduced the practice of covering them with poppy-petals, which service the Court of Directors of the East India Company acknowledged by presenting him with 50,000 rupees. Sometimes these balls are so soft as to burst their skins, when much of the liquid opium is lost. The quantity of opium produced annually in Bengal exceeds five millions of pounds, and the income derived by the Hon. East India Company from this source is not less than £5,003,162.

The kinds of opium most approved in the English market is the Smyrna, and in China and the East generally, the preference is given to the produce of India. Before used by the opium-smoker, the extract undergoes a course of preparation, the following being the method pursued in Singapore, as described by Mr. Little.

Between three and four o'clock in the morning the fires are lighted. A chest is then opened by one of the officers of the establishment of the opium farmer, and the number of balls delivered to the workmen proportioned to the demand. The balls are then divided into equal halves by one man, who scoops out with his fingers the inside or soft part, and throws it into an earthen dish, frequently during the operation moistening and washing his hands in another vessel, the water of which is carefully preserved. When all the soft part is carefully abstracted from the hardened skins or husks, these are broken up, split, divided, and torn, and thrown into the earthen vessel, containing the water already spoken of, saving the extreme outsides, which are not mixed with the others, but thrown away, or sometimes sold to adulterate chandu in Johore and the back of the island.

The second operation is to boil the husks with a sufficient quantity of water in a large, shallow, iron pot, for such a length of time as may be requisite to break down thoroughly the husks, and dissolve the opium. This is then strained through folds of China paper, laid on a frame of basket-work, and over the paper is placed a cloth. The strained fluid is then mixed with the opium scooped out in the first operation, and placed in a large iron pot, when it is boiled down to the consistence of thickish treacle. In this second operation, the refuse from the straining of the boiled husk is again boiled in water, filtered through paper, and the filtered fluid added to the mass, to be made into chandu. The refuse is thrown outside, and little attended to. It is dried and sold to the Chinese going to China for from ten to seventeen shillings the hundredweight, who pound it, and adulterate good opium with it. The paper that has been used in straining contains a small quantity of opium, it is carefully dried and used medicinally by the Chinese.

In the third operation, the dissolved opium being reduced to the consistence of treacle, is seethed over a fire of charcoal, of a strong and steady, but not fierce temperature, during which time it is most carefully worked, then spread out, then worked up again and again by the superintending workman, so as to expel the water, and, at the same time, avoid burning it. When it is brought to the proper consistence, it is divided into half-a-dozen lots, each of which is spread like a plaister on a nearly flat iron pot, to the depth of from half to three-quarters of an inch, and then scored in all manner of directions to allow the heat to be applied equally to every part. One pot after another is then placed over the fire, turning rapidly round, then reversed, so as to expose the opium itself to the full heat of the red fire. This is repeated three times, the length of time requisite, and the proper heat are judged of by the workman, from the effluvium and the colour, and here the greatest dexterity is requisite, for a little more fire, or a little less would destroy the morning's work, or eighty or a hundred pounds' worth of opium. The head workmen are men who have learned their trade in China, and from their great experience, receive high wages.

The fourth operation consists in again dissolving this fired opium in a large quantity of water, and boiling it in copper vessels till it is reduced to the consistence of the chandu used in the shops. The degree of tenacity being the index of its complete preparation, which is judged of by drawing it out with slips of bamboo.

By this long process, many of the impurities in the opium are got rid of, and are left in the refuse thrown out, such as vegetable matter, part of the resin and oil, with the extractive matter. By the seething process, the oil and resin are almost entirely dissipated, so that the chandu, as compared with the crude opium, is less irritating and more soporific. The quantity of chandu obtained from the soft opium is about seventy-five per cent., but from the opium, including the husk, not more than 50 to 54 per cent.

The heat to be endured by the men during this operation is very great, and can only be tolerated when custom has inured them to it. One of these men, Mr. Little graphically describes. He was quite a character in his way.

> From three in the morning till ten in the forenoon he stands before the boiling cauldron, with a fan in one hand, and a feather in the other; with the latter he scoops off the scum that forms, while, with the fan, he prevents the fluid from boiling over. He never speaks, but is always smiling; nor does he move, except to quench his thirst, from a bucket of water placed beside him. His trowsers are his only article of dress, the floor his bed, a little rice his food. When his labour is finished, his enjoyment is to drink arrack till he is insensible, from which he is wakened in the morning to his work. He has but one idea, and that is, the prospect of getting drunk on his favourite beverage; for his work is mechanically done, and costs him not a thought, no more than it does the dog that turns the spit. But he smiles, as he thinks of the revel for the night; and with his whole soul wrapped up in that fancied bliss, he heeds not the days that go by. He is a singular being, and in another country, would be the inmate of a mad-house.

The method of preparation in China and Hong-Kong, is identical with that pursued at Singapore. When the chandu or prepared extract of opium is consumed, it leaves a refuse consisting of charcoal, empyreumatic oil, some of the salts of the opium, and part of the chandu not consumed. One ounce of the chandu gives nearly half an ounce of the refuse called *Tye* or *Tinco*. This is smoked or swallowed by the poorer classes, who cannot afford the pure extract, and for this they only pay half the price of chandu. When smoked, it yields a further refuse called *Samshing*, which contains a very small quantity of the narcotic principle. This last is never smoked, as it cannot furnish any smoke, but is swallowed, and that not unfrequently mixed with arrack. Samshing is used by the very poorest and most indigent class—by beggars and outcasts, and those who, from long habit, are unable to exist without some stimulus from the drug, but are unable to supply themselves with any but the cheapest form in which any of the effects of the narcotic can be obtained.

Opium is called in Arabic "Afiyoon," and the opium-eater "Afiyoonee." In the crude state, opium is generally taken by those who have not long been addicted to its use, in the dose of three or four grains, and the dose is increased by degrees.

The Egyptians make several conserves composed of hellebore, hemp, and opium, and several aromatic drugs which are in much more common use than simple opium. One of these conserves is called "magoon," and the person who makes or sells it, is called "magoongee." The most common kind is called "barsh" or "berch." There is one kind which, it is said, makes the person who takes it manifest his pleasure by singing, another which will make him chatter, a third which excites to dance, a fourth which particularly effects the vision in a pleasurable manner, and a fifth which is simply of a sedative nature. These are sold at certain kind of shops called "mahsheshehs," solely appropriated to the sale of intoxicating preparations.

Thus, in different countries, we find opium used in different ways. In Great Britain, for instance, it is either used in the solid state, made into pills, in which form it is somewhat extensively employed in certain of our manufacturing districts, where druggists are affirmed to keep a supply of these pills ready made to meet the demand, or it is used in the form of tincture in the common state of laudanum, in which form it is not only used medicinally, but to our knowledge, somewhat largely as a means of indulgence, or, we should rather say, with somewhat of qualification, largely for a country in which many are fain to suppose that it is not used for those purposes at all. It is also

used in the form of Paregoric elixir, and is given insiduously to children under a variety of quack forms, such as Godfrey's cordial, etc. On
the authority of a reverend gentleman, it is stated that in the town of
Preston, in 1843, there were upwards of sixteen hundred families in
which Godfrey's cordial was habitually employed, or some other equally
injurious compound. Professor Johnston has noticed a communication which appeared in the *Morning Chronicle*, describing the effects
of opium upon the health of children, says—"The child sinks into a
low torpid state, wastes away into a skeleton, except the stomach,
producing what is known as pot-belly. One woman said, 'The sleeping
stuff made them that they were always dozing, and never cared for
food. They pined away; their heads got big, and they died.'"

In India, the pure opium is either dissolved in water, and so used,
or rolled into pills. It is there a common practice to give it to children
when very young, by mothers who require to work, and cannot at the
same time nurse their offspring. The natives of the western coast of
Africa have a curious mechanical contrivance, by means of which
they get rid of the necessity for opium in these cases. The girls wear a
"kankey," or artificial hump on their backs as soon as they can walk,
in order to learn betimes to carry their juniors, who ride astride on the
said projections. The usefulness of them consists in enabling the mothers to work with their infants in this way *on their backs*, while in England they excuse themselves from work on the plea of an infant *in
arms*, or else the helpless little creatures are drugged with sleeping
stuff, and their heads grow big, and they die.

In China, opium is either swallowed or smoked in the shape of
Tye. In Bally it is first adulterated with China paper, and then rolled
up with the fibres of a particular kind of plantain. It is then inserted
into a hole made at the end of a small bamboo and smoked. In Java
and Sumatra it is often mixed with sugar and the ripe fruit of the
plantain. In Turkey it is usually taken in pills, and those who do so,
avoid drinking any water after having swallowed them, as this is said
to produce violent colic; but to make it more palatable, it is sometimes mixed with syrups or thickened juice; in this form, however, it is
less intoxicating, and resembles mead. It is then taken with a spoon,
or is dried in small cakes, with the words "Mash Allah," the "Work of
God," or the "Gift of God" imprinted on them. When the dose of two
or three drams a day no longer produces the beatific intoxication so

eagerly sought, they mix corrosive sublimate with the opium till the quantity reaches ten grains a day.

In Singapore there are representatives of almost every Eastern nation, indulging in the luxury according to the fashion of the country of which he is native. The Hindoo, fresh from the continent, prefers the mode there in use, and swallows the soul-soothing pill; while the Chinese, with a gusto which no worshipper of the meerschaum can compete with, inhales the smoke, not only into his mouth, but into his lungs, where it becomes breath of his breath, and where retained, it acts on the nervous fibres that are spread over the extensive membrane which lines every cell of the lungs until exhaled through nose and mouth—yea, even in some cases, through ear and eye, it is replaced by another puff.

As the body becomes accustomed by habit to bear larger doses of opium than before the habit has been formed, the enormous quantity which some persons have taken are startling and surprising. Dr. Christison, in his work on poisons, refers to some of these cases.

A female who died of consumption at the age of forty-two, had taken about a dram of solid opium daily for ten years. A well-known literary character, about fifty years of age, has taken laudanum for twenty-five years, with occasional short intermissions, and sometimes an enormous quantity, but enjoys tolerable bodily health. A lady about fifty-five, who enjoys good health, has taken opium many years, and at present uses three ounces of laudanum daily. Lord Mar, after using laudanum for thirty years, at times to the amount of two or three ounces daily, died at the age of fifty-seven, of jaundice and dropsy. A woman who had been in the practice of taking about two ounces of laudanum daily for very many years, died at the age of sixty or upwards. An eminent literary character who died lately, about the age of sixty-three, was in the practice of drinking laudanum to excess from the age of fifteen, and his daily allowance was sometimes a quart of mixture consisting of three parts laudanum and one of alcohol. A lady now alive, at the age of seventy-four, has taken laudanum in the quantity of half an ounce daily between thirty and forty years. An old woman died not long ago at Leith at the age of eighty, who had taken about half an ounce of laudanum daily for nearly forty years, and enjoyed tolerable health all the time. Visrajee, a celebrated Cutchee chief mentioned by Dr. Burnes, had taken opium largely all his life, and was alive at the age of eighty, with his mind unimpaired.

To these examples we may add the confession of De Quincey:

> I, who have taken happiness both in a solid and a liquid shape, both boiled and unboiled, both East Indian and Turkish—who have conducted my experiments upon this interesting subject with a sort of galvanic battery, and have, for the general benefit of the word, inoculated myself, as it were, with the poison of eight thousand drops of laudanum a day—I, it will be admitted, must surely now know what happiness is, if anybody does. Fifty and two years' experience of opium, as a magical resource under all modes of bodily suffering, I may now claim to have had. According to the modern slang phrase, I had, in the meridian stage of my opium career, used 'fabulous' quantities. Stating the quantities—not in solid opium, but in the tincture (known to everybody as laudanum)—my daily ration was eight thousand drops. If you write down that amount in the ordinary way as 8000, you see at a glance that you may read it into eight quantities of a thousand, or into eight hundred quantities of ten; or, lastly, into eighty quantities of one hundred. Now, a single quantity of one hundred will about fill a very old-fashioned obsolete teaspoon, of that order which you find still lingering amongst the respectable poor. Eighty such quantities, therefore, would have filled eighty of such antediluvian spoons, that is, it would have been the common hospital dose for three hundred and twenty adult patients.

And he adds solemnly, that "without opium, thirty-five years ago, beyond all doubt, I should have been in my grave."

It is not a very easy task to ascertain the full extent of opium indulgence at home; but there is more of truth than fiction in that passage in *Alton Locke*, where the hero, on his way to Cambridge, meets with a ride in the vehicle of a certain yeoman of the Fen country, and enters into conversation with him, in the course of which the following dialogue takes place.

"Love ye, then! they as dinnot tak' spirits down thor, tak' their pennord o' elevation, then—women folk especial."

"What's elevation?"

"Oh! ho! ho! Yow goo into druggist's shop o' market day, into Cambridge, and you'll see the little boxes, doozens and doozens, a' ready on the counter; and never a ven-man's wife goo by, but what calls in for her pennord o' elevation, to last her out the week. Oh! ho! ho! Well, it keeps women folk quiet, it do; and it's mortal good agin ago pains."

"But what is it?"

"Opium, bor' alive, opium!"

"But doesn't it ruin their health? I should think it the very worst sort of drunkenness."

"Ow, well, yow moi say that—mak'th 'em cruel thin, then, it do; but what can bodies do i' th' ago? But it's a bad thing, it is."

The fact is well known, that in the Fen country, opium is extensively used under the presumption or excuse that it is good for the ague. In Wisbeach, as we ascertained from certain official documents, more opium is sold and consumed, in proportion to population, than in any other part of the kingdom. In other parts of Cambridgeshire and Lincolnshire, large quantities of opium are regularly and habitually sold in small doses amongst the labouring population. In Manchester some years ago, a similar run upon opium was experienced, but *not* as a cure for ague. Several cotton manufacturers stated to our authority, that their work-people were rapidly getting into the practice of opium-eating; so much so, that on a Saturday afternoon the counters of the druggists were strewed with pills of one, two, or three grains, in preparation for the known demand of the evening. The immediate occasion of this practice was stated to be the lowness of wages, which, at that time, would not allow them to indulge in ale or spirits; hence they adopted opium as a substitute.

There was a sin of which we were guilty in the age of Butler, and from which we are not yet freed; probably, it is somewhat of a universal one. Whether or no, there are certainly not a few who—

Compound for sins they are inclined to,
By damning those they have no mind to.

Opium indulgence is, after all, very un-English, and never has been, nor ever will be, remarkably popular; and if we smoke our pipes of tobacco ourselves, while in the midst of the clouds, we cannot forbear expressing our astonishment at the Chinese and others who indulge in opium. Pity them we may, perhaps, looking upon them as miserable wretches the while, but they do not obtain our sympathies. Philanthropists at crowded assemblies denounce, in no measured terms, "the iniquities of the opium trade," and then go home to their pipe or cigar, thinking them perfectly legitimate, whether the produce of slave labour or free. It is the same sort of feeling that the Hashasheens of the East inspire, and indeed all, who have a predilection for other narcotics than those which Johnny Englishman delights in, come in for a share of his contempt.

A carrion crow was once indulging in a feast upon the carcase of a nice fat rat which had just been caught in a neighbouring barn and thrown out into the road. A wood pigeon, who had finished his meal in a field of peas hard by, came past at the time and saw his friend the crow in full enjoyment of his rat. "I cannot imagine," said the pigeon, "how you can eat such a disgusting creature as that on which you are making your breakfast—the sight of it turns my stomach." "It is quite a matter of taste," said the crow, "and I think that I have the advantage, my food is juicy and sweet, this rat has lived upon the best of the farmer's corn, and the farmer would enjoy the treat himself, I am confident, if he only knew what a delicious breakfast it would make. You should be welcome to an acre of peas every day, if you would bring me such a dish as this. Besides, if I did not eat it, it would soon putrefy, and fill the air with disgusting smells, so that I am, in myself, a perfect board of health, working for the good of society, you, no better than a vagabond, stealing from society your daily bread." "I have heard it said," added the pigeon, "that it was you and your companions that destroyed a whole field of turnips in grubbing after the worms—I suppose that was a benefit to society." "Go and eat your peas," said the crow, "and leave me to enjoy my rat in peace."

Calculations as to the number of persons indulging in the use of opium are necessarily liable to objections; one person asserting that in China, for instance, not less than twenty millions of people indulge in opium, whilst others consider that two millions and a half are all that can be calculated upon. The number which Johnston estimates as the proportion of the human race using opium is four hundred millions, or about half the number of those who indulge in tobacco. This is, perhaps, as near an approximation as can be made, but one which must be based on the quantity produced, deducing therefrom the number required to consume it, rather than on any details of consumption, which cannot be arrived at.

There is one important and well-authenticated fact with regard to the Chinese consumption of opium, that in the year 1854, the value of opium imported into China exceeded the value of all the tea and silk exported from China to Great Britain and her colonies.

As we take farewell of the "gift of God" to pass through the portals of Paradise, let us do so in the words of that most celebrated of English opium-eaters, Thomas de Quincey:

> O just, subtle, and all-conquering opium! that, to the hearts of rich and poor alike, for the wounds that will never heal, and for the pangs of grief that "tempt the spirit to rebel," bringest an assuaging balm; eloquent opium! that with thy potent rhetoric stealest away the purposes of wrath, pleadest effectually for relenting pity, and through one night's heavenly sleep, callest back to the guilty man the visions of his infancy, and hands washed pure from blood. O just and righteous opium! that to the chancery of dreams, summonest for the triumphs of despairing innocence, false witnesses, and confoundest perjury, and dost reverse the sentences of unrighteous judges; thou buildest upon the bosom of darkness, out of the fantastic imagery of the brain, cities and temples, beyond the art of Phidias and Praxiteles—beyond the splendours of Babylon and Hekatompylos; and from the "anarchy of dreaming sleep," callest into sunny light the faces of long-buried beauties, and the blessed household countenances, cleansed from the "dishonours of the grave." Thou only givest these gifts to man, and thou hast the keys of Paradise, O just, subtle, and mighty opium!

CHAPTER X

THE GATES OF PARADISE

Thou only givest these gifts to man; and thou hast the keys of
Paradise, O just, subtle, and mighty opium.
                                    Confessions of an Opium-Eater

According to the common opinion of the Arabs, there are seven heavens, one above the other. The upper surface of each is believed to be nearly plane, and generally supposed to be circular, five hundred years' journey in width. The first is described to be formed of emerald; the second of white silver; the third of large white pearls; the fourth of ruby; the fifth of red gold; the sixth of yellow jacinth; and the seventh of shining light. Some assert Paradise to be in the seventh heaven; others state that above the seventh heaven are seven seas of light, then an undefined number of veils, or separations, of different substances, seven of each kind, and then Paradise, which consists of seven stages, one above another. The first is the mansion of glory, of white pearls; the second, the mansion of peace, of ruby; the third, the garden of rest, of green chrysolite; the fourth, the garden of eternity, of green coral; the fifth, garden of delight, of white silver; the sixth, the garden of Paradise, of red gold; the seventh, the garden of perpetual abode or Eden, of large pearls—this overlooking all the former, and canopied by the throne of the Compassionate.

The most direct road and speediest conveyance to Paradise, according to the testimony of all confirmed opiophagi, is by means of that subtle drug, opium. The most common form in which it is taken is that of vapour, inhaled through a peculiarly-constructed pipe. Those used by the Siamese resemble in form the common narghilè, or hubble-bubble of the Levant. They consist of an empty cocoa-nut shell, in an orifice in the top of which a hollow wooden tube is inserted, and the opening hermetically closed, so as to prevent the escape of either air or smoke. In another hole in the side of the cocoa-nut shell, a common little bamboo tube, about eighteen inches long, is tightly fixed; a little earthen bowl, perforated at the bottom like a sieve, is filled with opium, and one or two pieces of fire being placed thereon, this bowl is fitted on the top of the wooden tube. The man who hands round this pipe holds with one hand the bottom of the cocoa-nut (which is half full of water), and with the other hand he presents the bamboo tube to the smoker, who, putting it to his mouth, inhales three or four whiffs of this most intoxicating narcotic. The effect is almost instantaneous. He sinks gently against the cushion set at his back, and becomes insensible to what is passing around. The pipe is passed round from mouth to mouth, so that half an hour generally intervenes between the first whiff taken by the first smoker, and the last sigh heaved by the last man, as he revives from his short, pleasant dream, into which the whiffing has thrown him. One old and inveterate Siamese smoker declared to a recent resident among them, that if he knew his life would be forfeited by the act, he could no more resist the temptation than he could curb a fiery steed by a thread bridle. It carried him into the seventh heaven—he heard and saw things no tongue could utter, and felt as though his soul soared so high above things earthly, during those precious moments of oblivion, as to have flown beyond the reach of its heavy, burthensome cage.

Opium-smoking is not generally conducted on a plan so social. The Siamese may be considered as an exception to the general rule. The method pursued at Hong-Kong, of which we have received an account from a competent authority, is more a type of the opium-smoker in general, and the method he pursues.

In a reclining position, on boards placed on tressels, ranged around long, disgustingly dirty rooms, may be seen, at all hours of the day, haggard beggars, with putrefying sores, whose miserable feelings of desperation and woe drive them here to obtain a partial alleviation, by steeping their senses in forgetfulness. The stem of the pipe used for smoking is made of hard wood, and would be taken for an English paper-ruler, about eighteen inches long, and an inch in diameter. The earthenware bowl or head screws on and off, at about three inches from the end. An assistant of the divan, sitting in a corner of the room, is constantly engaged in scraping and cleaning these heads, which, from the small size of the hole through which the opium is inhaled (about the size of a pin's head), are apt to get clogged. The quantity of opium intended to be smoked, varying at a time from twenty to a hundred grains, is dipped carefully out of small gallipots, laid on a leaf, and charged for at the rate of a dollar per ounce. The opium is used by dipping into it the pointed end of a small wire, which is then applied to the flame of a lamp. In ignition it inflates into a bubble, and is then, with a dexterity obtained only by constant practice, rolled on the pipe head until it assumes the shape and size of a small orange-pip cut in half, and the hardness of wax. It is then placed over the orifice in the head of the pipe, like a small chimney, through which the flame of the lamp is drawn into the bowl, converting the opium, in its passage, into a blue smoke, which is inspired by long continuous whiffs, and without removal of the pipe from the mouth, respired through the nostrils. Two or three pipes may be taken by persons un-accustomed to the habit without leaving any other unpleasant feeling than a harshness in the throat. There are in Hong-Kong ten regular licensed divans for the smoking of opium, and nearly all these are in the Chinese portion of the town.

This picture would, however, be incomplete, without a few more particulars concerning the individuals who give themselves up to indulgence in the drug. And for this we must again seek the aid of an experienced medical man, who for years lived and laboured in the midst of opium-smokers. "Nothing on earth," he states,

> can equal the apparent quiet enjoyment of the opium smoker. As he enters the miserable scene of his future ecstasy, he collects his small change, the labour, or begging, or theft of the day, with which he supplies himself with his quantity of Chandu; then taking the pipe, which is furnished gratis, he reclines on a board covered with a mat, and with his head resting on a wooden or bamboo pillow, he commences filling his pipe. As he entered, his looks were the picture of misery, his eyes were sunk, his gait slouched, his step trembling, and his voice quivering, with a sallow cast of countenance,

and a dull unimpressive eye. He who runs might read that he is an opium-smoker, and, diving still deeper below appearances, would declare him an opium sufferer. But now with pipe in hand, opium by his side, and a lamp before him, his eye already glistens, and his features soften in their expression, while he is preparing the coming luxury. At last it is ready, and the pipe being applied to the lamp, there is heard a soughing noise, as with a full and hearty pull, he draws in all that opium and air can give. Slowly is the inspiration relaxed, but not until all the opium that is in the pipe is consumed; then, allowing the vapour, impregnated with the narcotic influence, to remain in his chest until nature compels him to respire, he gently allows it to escape, seeming to grudge the loss of each successive exit, until all is gone, when exhausted and soothed—

> Like one who wraps the drapery of his couch
> About him, and lies down to pleasant dreams,

he withdraws the pipe, reclines his head, and gives himself up to the first calming effect of the drug. His next attempt confirms the comfort, and now no longer does he complain of racking limbs or aching bones; no longer does the rheum run from his eyes, and relaxed is the tightness of the chest, as he dwells with fond affection on the inspiring pipe. His second pipe being finished, he can now look round, and has time to gaze on what is going on; but his soul is still wrapped in the bliss that is anticipated from what remains of his allowance, for not until a third or fourth whiff do the feelings of positive pleasure arise. Then is felt a lightness of the head, a tingling in every limb—the eyes seem to be enlarged, and the ears sharpened to hearing, an elasticity, an inclination to mount on high is experienced—all pains are gone, and pleasure now remains—all weariness has left, and freshness takes its place. The loathing of food that was lately experienced is changed to a relish for what is piquant, and a great desire is frequently felt for some particular food. The tongue is now loosened and tells its tale. For whatever is secret becomes open, and what was intended for one becomes known to all. Still there is no excitement, but a calmness, soft, soothing, and sedative. He dreams no dreams, nor thinks of the morrow but with a smile in his eye; he fills his pipe with the last of his allowance;

slowly inhaling it, he seems to brighten up. The smile that was sparkling in his eye, extends to other features, and his appearance is
one of complete, yet placid enjoyment. Presently the pipe is slowly
displaced, or drops by his side; his head, if raised, is now laid on the
pillow—feature after feature gives up its smile—the eye becomes
glazed—now droops the upper eyelid, and falls the chin with the
lower lip, deeper and deeper inspirations follow—all perception is
gone; objects may strike the eye, but no sights are seen; sounds may
fall on the ear, but no sensations are excited. So he passes into sleep,
disturbed and broken, from which the wretched being awakes to a
full conception of his misery. 'To sleep, perchance to dream!'—and
what dreams!—what ecstatic delights!—what ravishments!—what
illusions!

> Things
> Seen for the first time, and things, long ago
> Seen, which he ne'er again shall see, do blend
> Strangely and brokenly with ghastly things
> Such as we hear in childhood, scorn in youth
> And doubt in manhood, save when seen.

In the narrative of the voyage of H.M.S. *Samarang*, Mr. A. Adams
informs us, that in a large caravansary belonging to the Malay village
near Singapore, he had an opportunity of observing the effects of opium
on the physical aspect of the Malay. One of these was a feeble, worn
out old man, with an unearthly brilliancy in his eye. His body was
bent forwards and greatly emaciated; his face was shrunken, wan, and
haggard; his long skinny arm, wasted fingers, and sharp pointed nails
resembled more the claw of some rapacious bird, than the hand of a
lord of the creation; his head was nodding and tremulous; his skin
wrinkled and yellow; and his teeth were a few decayed, pointed, and
black stained fangs. As he was approached, he raised his body from
the mat on which he was reposing. There was something interesting
and at the same time melancholy in the physique of the old man, who
now in rags, appeared from the silver ornaments he wore, and by his
embroidered jacket, to have been formerly a person of some distinction; but the fascinating influence of the deadly drug had fastened on
him, and a pallet in a caravansary was the reward of self-indulgence.

"In my experience of opium," says Mr.—,

which has not, however, been very extensive, I cannot say I have
found as much pleasure as the English opium-eater in his *Confessions* would lead us to believe fell to his lot. After three or four
Chinese opium-pipes, I found my brain very much unsettled, and
teeming with thoughts ill-arranged, and pursuing each other in
wanton dreamy play, without order or connection, the circulating
system being at the time much excited, the frame tremulous, the
eyeballs fixed, and a peculiar and agreeable thrilling sensation extending along the nerves. The same succession of image crowding
upon image, and thoughts revelling in strange disorder, continues
for some time, during which a person appears to be in the condition
of the madman alluded to by Dryden in his play of the *Spanish Fryar*.

> He raves, his words are loose,
> As heaps of sand, and scattering wide from sense
> So high he's mounted on his airy throne,
> That now the wind has got into his head,
> And turned his brains to frenzy.

Unutterable melancholy feelings succeed to this somewhat pleasurable period of excitement, but a soft languor steals shortly across
the senses, and the half-poisoned individual falls asleep. The next
day there is great nausea and sickness of stomach, headache, and
tormenting thirst, which makes you curse opium, and exclaim, with
Shakespeare's *King John*,

> And none of you will bid the winter come
> To thrust his icy fingers in my maw;
> Nor let my kingdom's rivers take their course
> Thro' my burnt bosom, nor entreat the North
> To make his bleak winds kiss my parched lips,
> And comfort me with cold.

Dr. Madden tried, experimentally, the effects of opium—he commenced with a grain, which produced no perceptible effect, to this he
afterwards added another grain. After two hours from commencing
the operation, his spirits became excited. "My faculties," he writes,

appeared enlarged, everything I looked at seemed increased in volume. I had no longer the same pleasure when I closed my eyes which

I had when they were open; it appeared to me as if it was only external objects which were acted on by the imagination, and magnified into images of pleasure; in short, it was the faint exquisite music of a dream in a waking moment. I made my way home as fast as possible, dreading, at every step, that I should commit some extravagance. In walking, I was hardly sensible of my feet touching the ground—it seemed as if I slid along the street, impelled by some invisible agent, and that my blood was composed of some ethereal fluid, which rendered my body lighter than air. I got to bed the moment I reached home. The most extraordinary visions of delight filled my brain all night. In the morning I rose pale and dispirited, my head ached, my body was so debilitated, that I was obliged to remain on the sofa all day, dearly paying for my first essay at opium-eating.

Thus far, the opium-eater and the opium-smoker seem to agree in the principal results from the use of the drug.

From the communications of Dr. Medhurst may be learnt many important facts relative to this habit in China. Day by day, and year by year, the practice of opium-smoking prevails more and more among this people, and by and by it will doubtless have a powerful effect upon the destinies of the country. It is said, that the late Emperor used the drug; it is certain that most of the government officers do, and their innumerable attendants are in the same category. Opium is used as a luxury by all classes, and to a great extent, indeed so great, that it cannot fail to exhibit its effects speedily upon the mass of the inhabitants. In rich families, even if the head of the house does not use the drug, the sons soon learn to use it, and almost all are exposed to the temptation of employing it, as many of their friends and acquaintances are in the habit of smoking; and it is considered a mark of politeness to offer the pipe to a friend or visitor. Many persons fly to the use of the pipe when they get into trouble, and when they are afflicted with chronic or painful diseases, sleeplessness, etc. Several persons who have been attended for malignant tumours were made victims of the drug, by the use of it to appease the pain and distress they had to endure. The beggars are, to a great extent, under its influence; but they use the dregs and scrapings only of the half-consumed drug, which is removed from the pipe-head when it is cleaned. The most common cause of the Chinese resorting to the use of the opium-pipe is their not knowing how to employ their leisure hours when the business of the day is over—there is no periodical literature to engage their attention. Their families do not present sufficient attractions to keep them at home, and sauntering about of an evening, with nothing to employ the mind, they are easily tempted into the opium shops, where one acquaintance or another is sure to be found, who invites to the use of the drug.

Many of the middling classes dissipate their money in this indulgence, and, among the lower classes, those who indulge in the use of opium are reduced to abject poverty. Having no property, furniture, or clothes to dispose of, their wives and children are sold to supply their ever-increasing appetite for the drug; and when these are gone, with greatly diminished strength for labour, they can no longer earn sufficient for their own wants, and are obliged to beg for their daily bread. As to the supply of opium, they must depend on the scrapings of other men's pipes, and as soon as they are unable, by begging, to obtain the necessaries of life, together with the half-burnt opium, on which their very life depends, they droop and die by the roadside, and are buried at the expense of the charitable.

Two respectable young men, the sons of an officer of high rank, well informed, having received a good education, accustomed to good society, and who excited great interest in the minds of those with whom they came in contact, lately died. So inveterate was their habit of opium-smoking, and so large the quantity necessary to keep up the stimulus, that their funds were exhausted. Friends assisted them, and relieved their necessities again and again; but it was impossible to give them bread and opium too, and they subsequently died, one after the other, in the most abject and destitute condition.

At Shanghae, just inside the north gate, in front of a temple, one of such destitute persons, unable to procure either food or opium, was lying at the last gasp, while two or three others with drooping heads were sitting near, who looked as if they would soon be prostrated too. The next day, the first of the group lay dead and stiff, with a coarse mat wound round his body for a shroud. The rest were lying down unable to rise. The third day another was dead, and the remainder nearly so. Help was vain, and pity was the only feeling that could be indulged.

It may be judged of the extent of opium-smoking in China from the reports of the native Teapoas, inclosures in Sir J. Bowring's Report. The inhabitants in the Chung-wan (Centre bazaar) are about 5,800. The number that smoke opium, merely because they like it, are upwards of 2,600. In the Hah-wan (Canton bazaar) there are upwards of 1,200. The number that smoke opium, merely because they like it, are upwards of 600. At Sheong-wan the number of male residents are 13,000; there are 3,000 opium-smokers. At Tai-ping-shan the number of inhabitants are 5,300 men; of these upwards of 1,200 smoke opium because they like it. The number of inhabitants in Ting-loong-chow are 2,500; the number of opium-smokers are reported at 400. Thus, out of 27,800 inhabitants, 7,800 of whom, or 26 per cent., are smokers of opium.

Dr. McPherson, in writing of the Shikhs, informs us that most of the Shirdars are under the influence of spirits or of opium for eighteen hours out of the twenty-four. Their early use, both of the spirit and the drug, renders them indispensable through life. If deprived of their usual dose, the Shikh is one of the most wretched beings imaginable. Before engaging in any feast, the Shikh takes his opium, by which he is for a time excited, and this is soon followed by languor and inactivity. Talking of Runjeet Sing, who was at that time labouring under paralysis, from which eventually he died, he says he still used opium, so that little could be expected from remedial means.

The Shikhs are forbidden the use of tobacco by the tenets of their religion, but find a ready substitute for it in opium, which is consumed in great quantities throughout the whole of the Punjaub, as well as among the protected Shikh states. While under the effects of this drug, the Shikh is a very different person to the same individual before he has taken it. In the former instance, he is active and talkative; in the latter, lazy and stupid.

It has been imagined that the preparation of opium has an injurious effect upon those engaged therein; but Dr. Eatwell, of the Benares Agency, states that,

> amongst the thousands of individuals, cultivators, and *employés*, with whom the factory is filled during the receiving and manufacturing seasons, no complaints are ever heard of any injurious effects resulting from the influence of the drug, whilst they all remain quite as free from general sickness as persons unconnected with the general

establishment—in fact, if anything, more so. It occasionally happens that a casual visitor to the factory complains of giddiness or headache; but the European officers employed in the department, who pass the greater part of the day with the thermometer between 95° and 105° Fah. amongst tons of the drug, never experience any bad effects from it. The native purkhea sits usually from six A.M to three P.M. daily, with his hand and arm immersed nearly the whole time in the drug, which he is constantly smelling, and yet he feels no inconvenience from it. He has informed me, that at the commencement of the season, he experiences usually a sensation of numbness in the fingers; but I believe this to be more the result of fatigue, consequent upon the incessant use of the arm and fingers, than of any effect of the opium. In the large caking-vats, men are employed to wade knee-deep through the drug for several hours during the morning, and they remain standing in it during the greater part of the rest of the day, serving out the opium by armsful, their bodies being naked, with the exception of a cloth about the loins. These men complain of a sensation of drowsiness towards the end of their daily labours, and declare that they are overpowered early in the evening by sleep, but they do not complain of the effect as being either unpleasant or injurious.

Infants, of a few months old, may be frequently seen lying on the opium-besmeared floor, under the vats, in which dangerous position they are left by their thoughtless mothers; but, strange to say, without any accident ever occurring. Here are abundant facts to show that the health of those employed in the opium-factory, and in the manipulation of the drug, is not exposed to any risk, whatever; whilst the impunity with which the drug is handled by hundreds of individuals, for hours together, proves that it has no endemic action; for I am inclined to consider the soporific effect experienced by the vat-treaders as produced through the lungs, and not through the skin.

This may be considered, therefore, as setting the question entirely at rest, and demonstrating the fact that the factory labourers are not sufferers.

According to a Chinese petition presented on one occasion to the Emperor, it is believed that the English, by introducing opium into that country, did so as a means of its subjugation, presuming, we may suppose, that the Celestials were invincible, except by some such cabalistic means. "In the History of Formosa," says this document,

> we find the following passage: "Opium was first produced in Kaoutsinne, which by some is said to be the same as Kalapa or Batavia. The natives of this place were at the first sprightly and active, and, being good soldiers, were always successful in battle. But the people called Hung-maou (red-haired) came thither, and having manufactured opium, seduced some of the natives into the habit of smoking it. From these, the mania for it spread rapidly through the whole nation; so that in process of time the natives became feeble and enervated, submitted to foreign rule, and ultimately were completely subjugated. Now the English," it continues, "are of the race of foreigners called Hung-maou. In introducing opium into this country, their purpose has been to weaken and enfeeble the Central Empire. If not early aroused to a sense of our danger, we shall find ourselves, ere long, on the last step towards ruin."

The degradation or subjugation of the Chinese is much more likely to be affected by a habit concerning which we hear less, but which is infinitely more disastrous than the indulgence in opium. This is the brandy-drinking customs of the north. This horrible drink, distilled from millet, is the Chinaman's delight, and he swallows it like water. Many ruin themselves with brandy as others do with gaming. In company, or even alone, they will pass whole days and nights in drinking successive little cups of it, until their intoxication makes them incapable of carrying the cup to their lips. "Gambling and drunkenness," says Abbè Huc, "are the two permanent causes of pauperism in China."

It is unfortunately the custom for the distillers to supply brandy on credit for a whole year, so that a tippler may go on for a long time drawing from this inexhaustible spring. His troubles will only begin in the last moon—the legal period of payment. Then, indeed, he must pay, and with usury; and as money does not usually become more plentiful with a man from the habit of getting drunk every day, he has to sell his house and his land, if he has any, or to carry his furniture and his clothes to the pawnbroker's. In the south, there is less brandy-drinking, and more gambling; but between the two there is little to choose, as either impoverishes those who devote themselves to its service, and to which even opium-smoking is preferable.

Mr. Meadows, the Chinese Government Interpreter at Hong-Kong, says,

> As to the morality of the opium question, I am fortunately able to give the home reader, by analogy, and in a few words, as exact an idea of it as I have got myself. Smoking a little opium daily, is like taking a pint or two of ale, or a few glasses of wine daily; smoking more opium is like taking brandy as well as beer and wine, or a large allowance of these latter; smoking very much opium is like excessive brandy- and gin-drinking, leading to delirium tremens and premature death. After frequent consideration of the subject during thirteen years, the last two spent at home, I can only say that, although the substances are different, I can, as to the morality of producing, selling, and consuming them, see no difference at all; while the only difference I can observe in the consequences of consumption is, that the opium-smoker is not so violent, so maudlin, or so disgusting as the drunkard. The clothes and breath of the confirmed and constant smoker are more or less marked by the peculiar penetrating odour of opium, and he gets careless in time of washing from his hands the stains from his pipe. But all this is not more disagreeable than the beery, vinous, or ginny odour, and the want of cleanliness that characterize the confirmed drunkard. In all other respects, the contrast is to the disadvantage of the drunkard.

Without pursuing this question further, there is evidently a fascination in the pipe to the opium-smoker, to a degree of which the most ardent lover of a pipe of tobacco has but a faint idea. In proportion as the indulgence in the drug produces a state of happiness far transcending all that the votary of the weed experiences, so does its influence over him increase; and if it is difficult for the habitual smoker of tobacco to forego the pleasure of his accustomed pipe, it is therefore ten times more difficult for the smoker or eater of opium to renounce for ever, a custom which brings with it, even in imagination though it may be, tenfold more pleasures, and a more ecstatic enjoyment. This is the universal evidence of all who have been inquired of, and of all who have had intercourse with opiophagi in all parts of the world.

What fascinating influence this Paradise in prospect has upon those who indulge in journeys thither, may be imagined from the notorious fact, that in Bristol, Coleridge went so far as to hire men—porters, hackney-coachmen, and others—to oppose by force his entrance into any druggist's shop. But as the authority for stopping him was derived only from himself, so these poor men found himself in a fix; for when the time and the inclination arrived, he proceeded to the shop, and on offering resistance, he, the same who had instructed them to prevent his entrance, now insisted on their allowing him to pass, annulled all former instructions, and on the authority of one who paid for their services, demanded its exercise as he thought fit, and the gates of Paradise where opened.

According to Darwin, even poultry have mounted the ladder to within a few steps of Elysium; for that worthy informs us, that they were fed for the London market by mixing gin and opium with their food, and keeping them in the dark, but that "they must be killed as soon as they are fattened, or they become weak and emaciated, like human drunkards." We have no recording pullet to inform us of the visions of the barn-door family under the influence of the beatific drug, nor *Confessions of a Chanticleer*, to tell of the pains that succeeded a too-free indulgence in the little pills; all we learn from the account is, that the vision of Paradise very closely preceded its reality, for the feathered bipeds were dosed and killed. The human biped for half a century continues his dream—and all through that period it is but a dream—yet that he is happy while under its influence there can be no doubt; and when he has reclined on his couch, obtained his pipe, and sunk into the beatific oblivion so coveted by the Asiatic, we may imagine his exclaiming with the Peri, after obtaining the trickling tear,

> *Joy, joy for ever! my task is done;*
> *The gates are passed, and heaven is won.*
> *Oh! am I not happy? I am—I am.*
> *To thee, sweet Eden! how dark and sad*
> *Are the diamond turrets of Shadukram,*
> *And the fragrant bowers of Amberabad.*
> *Joy, joy for ever! my task is done;*
> *The gates are passed, and heaven is won!*

CHAPTER XI

REVELS AND REVERIES

That juice of earth, the bane
And blessing of man's heart, and brain—
That draught of sorcery, which brings
Phantoms of fair forbidden things
Whose drops, like those of rainbows, smile
Upon the mists that circle man
Brightening not only earth, the while
But grasping heaven, too, in their span.

Lalla Rookh

The Mahometan legend of their prophet's ascent into heaven, where he received instructions for the faith and conduct of his followers, is thus current amongst them.

As Mahomet was reclining on the sacred stone in the temple of Mecca, Gabriel came to him, and opened his breast from the breast-bone to the groin, and took out his heart, and washed it in a golden basin, full of the water of Faith, and then restored it to its place. Afterwards a white beast was brought to him, less than a mule, and larger than an ass, called Al-Borak. It had a human face, but the cheeks of a horse, its eyes were jacinths, and radiant as stars. It had eagle's wings, all glittering with rays of light, and its whole form was resplendent with gems and precious stones. Upon this Mahomet was borne. Gabriel proceeded with him to the first heaven of silver, and knocked at the door, after some conversation he was welcomed, and the door opened. Here Mahomet saluted Adam. They then proceeded to the second heaven, all of polished steel and dazzling splendour, and saluted Noah. They then entered the third heaven, studded with precious stones, and too brilliant for mortal eyes. Here was seen Azrael, the Angel of Death, writing continually in a book the names of those who are to be born, and blotting out those who are to die. They mounted to the

114

fourth heaven, of the finest silver, where they saw the Angel of Tears, who was appointed to weep over the sins of man, and predict the evils that awaited them. The fifth heaven was of purest gold. Here Mahomet was received and saluted by Aaron. This heaven was inhabited by the Avenging Angel. He sat on a throne surrounded by flames, and before him was a heap of red hot chains. The sixth heaven was composed of a transparent stone, where dwelt the guardian angel of heaven and earth. Here Moses wept at the sight of the prophet who was to have more followers than himself. Mahomet then entered the seventh heaven of divine light, where he saw many marvellous things, which he related for the instruction of the faithful. He entered Al Mamour, the house of Adoration, and as he entered, three vases were offered him, one containing wine, another milk, and a third honey. He drank of the milk, "Well hast thou done!" exclaimed Gabriel. "Hadst thou drunk of the wine, thy people had all gone astray." The Prophet then returned to earth, as he had ascended to heaven.

The Al-Borak of modern Moslems is opium, by means of this most miraculous of vehicles they mount to the heaven of heavens.

What are the true effects of opium are best described by an eminent physician, who has studied well the results produced by all such influences upon the brain. The imagination appears to be acted upon, independent of the peculiar torpor, accompanied by sensations of gratification, and the absence of all communication with the external world. The senses convey no false impressions to the brain; all that is seen, heard, or felt, is faithfully delineated, but the imagination clothes each object in its own fanciful garb. It exaggerates, it multiplies, it colours, it gives fantastic shapes; there is a new condition arising out of ordinary perception, and the reason, abandoning itself to the imagination, does not resist the delight of indulging in vision. If the eyes are closed, and nothing presented to excite the external senses, a whole train of vivid dreams are presented. A theatre is lighted up in the brain—graceful dancers perform the most captivating evolutions—music of an unearthly character floats along—poesy, whose harmonious numbers, and whose exciting themes, are far beyond the power of the human mind, is unceasingly poured forth. Memory is, however, generally asleep—all the passions, affections, and motions have lost their sway. It is all an exquisite indolence, during which dreams spontaneously arise, brilliant, beautiful, and exhilarating. There is order, harmony, tranquillity. If a single object has been vividly impressed upon the eye, it is multiplied a thousand times by the imagination— vast processions pass him in his reveries in mournful pomp.

That this is the doctrine of the true church on the subject of opium, we may learn from De Quincey, of which church he acknowledges himself to be the Pope, and self-appointed *legate à latere* to all degrees of latitude and longitude.

> I often fell into such reveries after taking opium, and many a time it has happened to me on a summer night, when I have been seated at an open window, from which I could overlook the sea at a mile below me, and could, at the same time, command a view of some great town standing on a different radius of my circular prospect, but at nearly the same distance—that from sunset to sunrise, all through the hours of night, I have continued motionless, as if frozen, without consciousness of myself as an object anywise distinct from the multiform scene which I contemplated from above. Such a scene in all its elements was not unfrequently realised for me on the gentle eminence of Everton. Obliquely to the left, lay the many languaged town of Liverpool; obliquely to the right, the multitudinous sea. The scene itself was somewhat typical of what took place in such a reverie. The town of Liverpool represented the earth, with its sorrows and its graves left behind, yet not out of sight nor wholly forgotten. The ocean, in everlasting but gentle agitation, yet brooded over by dove-like calm, might not unfitly typify the mind, and the mood which then swayed it. For it seemed to me as if then first I stood at a distance aloof from the uproar of life, as if the tumult, the fever, and the strife were suspended; a respite were granted from the secret burdens of the heart, some sabbath of repose, some resting from human labours. Here were the hopes which blossom in the paths of life, reconciled with the peace which is in the grave; motions of the intellect as unwearied as the heavens, yet for all anxieties a halcyon calm; tranquillity that seemed no product of inertia, but as if resulting from mighty and equal antagonisms, infinite activities, infinite repose.

And now let us follow him to the Opera.

The late Duke of Norfolk used to say, "Next Monday wind and

weather permitting, I propose to be drunk"; and, in like manner, I used to fix beforehand how often, within a given time, when, and with what accessory circumstances of festal joy, I would commit a debauch of opium. This was seldom more than once in three weeks, for at that time I could not have ventured to call every day (as afterwards I did) for "a glass of laudanum negus, warm, and without sugar."

No: once in three weeks sufficed; and the time selected was either a Tuesday or a Saturday night, my reason for which was this— Tuesday and Saturday were for many years the regular nights of per- formance at the opera house, and there in those times Grassini sang, and her voice was delightful to me beyond all that I had ever heard. Thrilling was the pleasure with which almost always I heard her. Shivering with expectation I sat, when the time drew near for her golden epiphany, shivering I rose from my seat, incapable of rest, when that heavenly and harp-like voice sang its own victorious welcome in its prelusive *threttanelo–threttanelo*. The choruses were divine to hear; and, when Grassini appeared in some interlude, as she often did, and poured forth her passionate soul as Andromache at the tomb of Hector, etc., I question whether any Turk, of all that ever entered the paradise of opium-eaters, can have had half the pleasure I had. But, indeed, I honour the barbarians too much, by supposing them capable of any pleasures approaching to the intel- lectual ones of an Englishman. A chorus of elaborate harmony dis- played before me, as in a piece of arras work, the whole of my past life—not as if recalled by an act of memory, but as if present, and incarnated in the music; no longer painful to dwell upon, but the detail of its incidents removed, or blended in some hazy abstrac- tion, and its passions exalted, spiritualized, and sublimed. And over and above the music of the stage and the orchestra I had all around me, in the intervals of the performances, the music of the Italian language talked by Italian women—for the gallery was usually crowded with Italians—and I listened with a pleasure, such as that with which Weld, the traveller, lay and listened in Canada, to the sweet laughter of Indian women; for the less you understand a lan- guage, the more sensible you are to the melody or harshness of its sounds.

Let the reader who seeks to know of his other Saturday evenings' experiences, wandering about in the market-places, and threading the intricate mazes of bye-lanes and alleys, seek it in his *Confessions*.

An Englishman awaking one morning finds himself at Hong-Kong, in the midst of opium and opium-smokers. He is astonished that the Chinaman loves opium as he loves nothing else; he cannot think why his vitiated taste had not settled upon something nobler, why he does not take a fancy to British Brandy? But no! he loves opium. And a Parsee takes him to see the lions, and is so civil as to convey the stranger into his warehouse and open two chests of opium, that he may see the drug as it passes into commerce. Of these, the first consisted of balls, which he describes as of the size of a large apple dumpling, and when cut open the mass is found to be solid. The other was full of objects which a commander in the navy ordered his men to return to the owners of a captured junk, "Ar'nt you ashamed, my lads, to loot a lot of miserable Dutch cheeses?" The "Dutch cheeses" were Patna opium, worth about £5 each. Globes of thick dark jelly enclosed in a crust not unlike the rind of a cheese. The Parsee tapped one with a fragment of an iron fastening of a chest, and drew forth about a spoonful of the drug. It was not the opium which engaged the traveller's attention, it was the effect it produced upon the surrounding coolies. He had never before seen excitement in a Chinaman's face. He had seen them tried for their lives, and condemned to death. He had seen them test the long-suffering patience of Mr. Tudor Davies in the Hong-Kong police court, where that gentleman was daily engaged in laborious endeavours to extract truth out of conflicting lies. He had seen them laugh heartily at a gesture at a sing-song; and he once saw a witness grin with great delight, as he unexpectedly saw his most intimate friend, a tradesman of reputed wealth, among a crowd of prisoners in the dock. But these coolies, when they saw that opium opened their horizontal, slit-shaped eyes, till they grew round and starting, their limbs, so lax and limpid, when not in actual strain of labour, were stiff from excitement, every head was pressed forward, every hand seemed ready to clutch. There was a possibility that it would be put down upon the window-sill, near which the stranger and his Parsee friend were standing—and there could be seen the shadow of fingers ready to be slid in. It was almost certain that it would be thrown aside—and there was the grand hope of an opium debauch gratis, and this was the state of mind that hope created. And oh what raptures, what delights, what dreams! Already, in imagination, they revelled in scenes such as the wakeful eye of mortal man ne'er saw, and such as never did the mind of man conceive.

> *A paradise of vaulted bowers*
> *Lit by downward gazing flowers,*
> *And watery paths that wind between*
> *Wildernesses calm and green,*
> *Peopled by shapes too bright to see*
> *And rest, having beheld; somewhat like thee*
> *Which walk upon the sea, and chaunt melodiously.*

We cannot understand this fascination in which opium holds its devotee to its full extent; and yet, in some sort, the lover of tobacco, deprived of his pipe or quid, can in some sort understand it better than any other Englishman, the opiophagi excepted. Let the admirer of his weed be placed in circumstances wherein he cannot indulge in that luxury, and the inward longings for his cherished companion are akin to those of the smoker of opium without his drug. Some inveterate smokers of tobacco have been known to declare that they would rather forego their accustomed meal than their whiff; this they will sometimes profess, but this the opium devotee often accomplishes. Instances are far from rare of opium-smokers dying of starvation, having denied their bodies the sustenance they required, to procure their much loved chandu. Martyrs to their love of opium.

As opium is generally indulged in by the lower classes, in establishments called Opium Shops, otherwise Papan Mera, a word or two belongs to them. In Singapore, these shops are limited by the regulations to forty-five in town and six in the country. Each has a red board, which the vendor ought to hang up outside his shop, with the number thereon, as received from the opium farmer. Hence the name of Papan Mera, or "red board," and the shops are known by that name by all classes of natives. They are scattered in all directions over the island; and wherever a number of Chinese are congregated, there you have one or more. The farmer is most interested in the sale of opium, and the extension of shops, and of the trade. A man goes to him generally, either previously known or recommended, and says he wishes to open a Papan Mera; of course, the opium farmer wishes that he may do so,

and be successful, and vend plenty of opium, all the opium being purchased of the opium farmer, no one else being allowed to sell opium in
the island, and for which privilege he contracts annually with the
Government in a handsome sum. The man gets the red board, for
which he pays two shillings. If the limited number of forty-five is completed he does not require a board, but he is not refused the privilege
of opening a shop. In this case, he hangs a mat in the place of the
door, by which an opium shop is known to all, while the fact is announced by a Chinese inscription. Nothing is paid for a licence, no
securities are entered into, but the new man purchases of the farmer a
certain quantity of chandu, or prepared opium, and according to his
facilities for selling it so is the price. If the shop is to be opened in
town, where there are more customers, and if near to where Chinese
artificers abound, then he pays about eight shillings a tael ($1^1/_3$ oz.), or
at the rate of six shillings an ounce. If at a little distance, about five
shillings and sixpence an ounce. Still further from town, five shillings, then four shillings and sixpence. Nay, it even descends to a fraction beyond three shillings an ounce. The last is the sum paid by the
Nacodah of a Chinese junk, who takes a large quantity at a time, as
two-thirds of his crew are generally consumers, and the facility for
illicit consumption is great. The proprietors of the Papan Mera are
expected to retail it to their customers at a little above the price at
which they have purchased it. If in town, where they pay tenpence a
cheen or six shillings an ounce, then they charge elevenpence a cheen
or scarcely seven shillings an ounce, to those who come to buy or use
it on the premises. The opium farmer receives nothing from the owner
of the shop, except the money for his opium; the owner receives nothing from the farmer but the opium for his money, and sometimes a
discount of eight per cent. Nor do the opium-smokers pay more at the
shops for their opium than if they purchased it direct from the farmer.
How, then, does the owner of the "red board" manage to live? How
does he pay rent, sometimes to the extent of £2 or £3 per month? How
can he keep his wife, and the little "red boards," and one or two coolies? Ecce! He does all this on the refuse of the chandu, the Tye or
Tinco, sold to the poor.

On the Tinco and Samshing, the owners of many of the opium
shops almost entirely depend for their living. By their sale the rent is
paid, the family supported, and the servants kept. If a man sells three

taels, or three ounces and three-quarters of chandu a day, there will be
about half that quantity of Tinco, or one ounce and three-quarters,
this is the unconsumed refuse left in the pipe after smoking, and which
is the property of the owner of the Papan Mera, and from the consumption of this he gets a further refuse of little more than three-
quarters of an ounce, which is called *Samshing*. If he sells his Chandu
for twenty-five shillings, by his Tinco and Samshing he will realize
nearly twelve shillings and sixpence a day, and this is his income. Few,
however, *sell* so much, and fewer still *receive* as much.

The Papan Mera is of all kinds, from a hovel to a brick house of
two stories, for which £3 monthly is paid for rent. Generally speaking,
the luxury of the pipe is all that the smoker cares for, and all other
things, such as commodious apartments, elegant furniture, and proper
ventilation are disregarded. In some houses there are apartments beside those entered from the street. The police regulations ordain that
at nine P.M. all shall give up their pipes. But is the sound of the curfew
always heeded? "Sooner would the panting traveller, under a burning
sun, when hours have elapsed, since his parched lips were moistened,
dash from his mouth the goblet before his thirst was half quenched,
than the opium-smoker be the slave of time." If nine o'clock comes,
and he has not reached his climax, he then retires to an inner chamber, where, at ease and undisturbed, he may realize that enjoyment,
and consummate that bliss, of which the owner of "blue coat and bright
buttons" would deprive him. Thus he slips into Paradise whilst the
Peri and the "peeler" remain outside disconsolate.

Our Papan Mera man is a good man, and his wife is a good woman,
so we get a peep indoors, upstairs, behind the scenes, the apartment
where ladies are at home *de jure*, not being allowed perhaps to smoke
at home *de facto*. Of course, the general visitor has no admittance. In
the centre stands a large bed, sitting up thereon a female, her back
supported with cushions. She is young, she is fair—yea, passing fair,
and dressed in the habiliments of the flowery land. Near her stands a
table, on which are tea and sweetmeats. She, too, is a votary to the
drug; with dreamy eyes half closed, she draws in the inspiring vapour,
then sinks back upon the cushions, unconscious that we are gazing
upon her, her dark dishevelled tresses hanging over, but scarce con-
cealing the heaving bosom, the only sign of life.

Although there are supposed to be but forty-five licensed opium shops in Singapore town, there are upwards of eighty; wherever there are Chinese, there may also be found the Papan Mera. Certain trades are congregated together—you have carpenters in one street, blacksmiths in another, gold and silver smiths in a third, and so on. Amongst some trades, the habit of opium-smoking is more common than in others, the principal consumers will be found amongst carpenters, blacksmiths, barbers, huxsters, coolies, boatmen, gambier planters, and gardeners. Full eighty-five per cent. of the persons engaged in these callings are devoted to the drug. Shoemakers, tailors, and bakers, are generally less addicted to the habit; amongst the two first-named, not more than twenty per cent. are smokers. Wherever you have carpenters, blacksmiths, etc., in abundance, there will you have opium shops in abundance also. In many streets there are six of these shops. In one street there are twelve. In Canton Street there are eight houses, and two of them are licensed for opium. At Hong-Kong and at Canton, the same thing occurs. Certain streets are devoted to certain trades, and certain trades devoted to opium.

M. Abbè Huc communicates a few additional facts concerning opium in China. At present this country purchases annually of the English, opium to the amount of seven millions sterling; the traffic is contraband, but it is carried on along the whole coast to the Empire, and especially in the neighbourhood of the five ports which have been opened to Europeans. Large fine vessels, armed like ships of war, serve as depots to the English merchants, and the trade is protected, not only by the English Government, but also by the mandarins of the Chinese Empire. The law which forbids the smoking of opium under pain of death, has, indeed, never been repealed; but everybody smokes away quite at his ease notwithstanding. Pipes, lamps, and all the apparatus are sold publicly in every town, and the mandarins themselves are the first to violate the law, and give this bad example to the people, even in the courts of justice. During the whole of the Abbè's long journey through China, he met with but one tribunal where opium was not smoked openly and with impunity.

The Chinese prepare and smoke their opium lying down, sometimes on one side, sometimes on the other, saying that this is the most favourable position; and the smokers of distinction do not give themselves all the trouble of the operation, but have their pipes prepared for them.

For several years past some of the southern provinces have been actively engaged in the cultivation of the poppy, and the fabrication of opium. The English merchants confess that the Chinese product is of excellent quality, though inferior to that of Bengal; but the English opium suffers so much adulteration before it reaches the pipe of the smoker, that it is not in reality so good as what the Chinese themselves prepare. The latter, however, though delivered perfectly pure, is sold at a low price, and only consumed by smokers of the lowest class. That of the English, notwithstanding its adulteration, thus writes Abbè Huc, is very dear and reserved to smokers of distinction; a caprice which can only be accounted for from the vanity of the rich Chinese, who would think it beneath them to smoke opium of native production, and not of a ruinous price; that which comes from a long way off must evidently be preferable. It is very probable that the Chinese will soon cultivate the poppy on a large scale, and make at home all the opium necessary for their consumption. It is certain that the English cannot offer an equally good article at the same price; and, should the fashion alter, British India will suffer a great reverse in her Chinese opium trade. The Abbè makes reference to the increased consumption of opium in England, both in the liquid and solid form, the progress of which he characterises as alarming, and then concludes the subject with the following extraordinary paragraph: "Curious and instructive would it be, indeed, if we should one day see the English going to buy opium in the ports of China, and their ships bringing back from the Celestial Empire this deleterious stuff, to poison England. Well might we exclaim in such a case, 'Leave judgement to God.'"

CHAPTER XII

PANDEMONIUM

Sights of woe,
Regions of sorrow, doleful shades, where peace
And rest can never dwell, hope never comes,
That comes to all.

Milton

The night side of opium-eating and smoking must be seen, as well as the bright and sunny day, before we lavish upon it encomiums, such as some of its votaries have indulged in. There may be a paradise to which the Theriaki can rise, but there is also an abyss into which he may fall. Lord Macartney informs us that the Javanese, under an extraordinary dose of opium, become frantic as well as desperate. They acquire an artificial courage; and when suffering from misfortune and disappointment, not only stab the objects of their hate, but sally forth to attack in like manner every person they meet, till self-preservation renders it necessary to destroy them. As they run they shout *Amok, amok*, which means *kill, kill!* and hence the phrase *running a muck*. The practice of running amok is hardly known at Pinang or any of the three Straits settlements. Captain Low did not recollect more than two instances at that place, including Province Wellesley, within a period of seventeen years, and the last he had heard of, which took place on shore at Singapore, was many years ago. A man ran *amok*—or, as the Malays term it, *meng amok*. He had gambled deeply, it was said, and had killed one or more individuals of his family. He next dosed himself with opium and rushed through the streets with a drawn kris or dagger in his hand, and pursued by the police. Major Farquhar, the then resident, hearing the uproar, went out of his house, where the infuriated man, who was just about to pass it, dashed at him, and wounded him in the shoulder; but a sepoy, who was standing as sentry at the door, received the desperado on his bayonet at the same instant, and prevented a second blow.

Captain Beeckman was told of a Javanese who ran a muck in the streets of Batavia, and had killed several people, when he was met by a soldier, who ran him through with his pike. But such was the desperation of the infuriated man, that he pressed himself forward on the pike, until he got near enough to stab his adversary with a dagger, when both expired together.

But the worst pandemonium which those who indulge in opium suffer, is that of the mind. Opium retains at all time its power of exciting the imagination, provided sufficient doses are taken; but when it has been continued so long as to bring disease upon the constitution, the pleasurable feelings wear away, and are succeeded by others of a very different kind. Instead of disposing the mind to be happy, it acts upon it like the spell of a demon, and calls up phantoms of horror and disgust. The fancy, still as powerful, changes its direction. Formerly it clothed all objects with the light of heaven—now it invests them with the attributes of hell. Goblins, spectres, and every kind of distempered vision haunt the mind, peopling it with dreary and revolting imagery. The sleep is no longer cheered with its former sights of happiness. Frightful dreams usurp their place, till at last the person becomes the victim of an almost perpetual misery.

The truth of all this is acknowledged by De Quincey, when writing of the pains of opium. Almost every circumstance becomes transformed into the source of terror. Visions of the past are still present in dreams, but not surrounded by a halo of pleasure any longer. The outcast Ann and the wandering Malay come back to torment him with their continued presence. All this is told in language so graphic, that it would be almost criminal to attempt its description in any other. The Dream of Piranesi is cited as a type of those he now suffered:

> Many years ago, as I was looking over Piranesi's "Antiquities of Rome," Coleridge, then standing by, described to me a set of plates from that artist, called his "Dreams," and which record the scenery of his own visions during the delirium of a fever. Some of these represented vast

Gothic halls, on the floor of which stood mighty engines and machinery—wheels, cables, catapults, etc.—expressive of enormous power put forth, or resistance overcome. Creeping along the sides of the walls, you perceived a staircase; and upon this, groping his way upwards, was Piranesi himself. Follow the stairs a little farther, and you perceive them reaching an abrupt termination, without any balustrade, and allowing no step onwards to him who should reach the extremity, except into the depths below. Whatever is to become of poor Piranesi! At least, you suppose that his labours must now in some way terminate. But raise your eyes, and behold a second flight of stairs still higher, on which again Piranesi is perceived, by this time standing on the very brink of the abyss. Once again elevate your eye, and a still more aerial flight of stairs is descried; and there again, is the delirious Piranesi, busy on his aspiring labours; and so on, until the unfinished stairs and the hopeless Piranesi both are lost in the upper gloom of the hall. With the same power of endless growth and self-reproduction did my architecture proceed in dreams. In the early stage of the malady, the splendours of my dreams were, indeed, chiefly architectural, and I beheld such pomp of cities and palaces as never yet was beheld by the waking eye, unless in the clouds. From a great modern poet, I cite the part of a passage which describes as an appearance actually beheld in the clouds, what, in many of its circumstances, I saw frequently in sleep:

> The appearance, instantaneously disclosed,
> Was of a mighty city—boldly say
> A wilderness of building, sinking far
> And self-withdrawn into a wondrous depth,
> Far sinking into splendour without end!
> Fabric it seem'd of diamond and of gold,
> With alabaster domes and silver spires,
> And blazing terrace upon terrace, high
> Uplifted; here, serene pavilions bright,
> In avenues disposed; there, towers begirt
> With battlements, that on their restless fronts
> Bore stars—illumination of all gems!
> By earthly nature had the effect been wrought
> Upon the dark materials of the storm

> *Now pacified; on them, and on the coves*
> *And mountain-steeps and summits, whereunto*
> *The vapours had receded—taking there*
> *Their station under a cerulean sky.*

Further confessions describe the characteristics of some of these opiatic visions in connection with tropical lands.

Under the connecting feeling of tropical heat and vertical sunlights, I brought together all creatures—birds, beasts, reptiles; all trees and plants, usages and appearances, that are found in all tropical regions, and assembled them together in China or Hindostan. From kindred feelings, I brought Egypt and her gods under the same law. I was stared at, hooted at, grinned at, chattered at by monkeys, by paroquets, by cockatoos. I ran into pagodas, and was fixed for centuries at the summit, or in secret rooms. I was the idol—I was the priest—I was worshipped—I was sacrificed. I fled from the wrath of Brama through all the forests of Asia—Vishnu hated me—Seeva lay in wait for me. I came suddenly upon Isis and Osiris. I had done a deed, they said, which the ibis and the crocodile trembled at. Thousands of years I lived, and was buried in stone coffins, with mummies and sphinxes, in narrow chambers, at the heart of eternal pyramids. I was kissed with cancerous kisses by crocodiles, and was laid, confounded with all unutterable abortions, amongst reeds and Nilotic mud.

Again he says:

The cursed crocodile became to me the object of more horror than all the rest. I was compelled to live with him, and (as was always the case in my dreams) for centuries. Sometimes I escaped, and found myself in Chinese houses. All the feet of the tables, sofas, etc., soon became instinct with life; the abominable head of the crocodile, and his leering eyes, looked out at me, multiplied into ten thousand repetitions; and I stood loathing and fascinated. So often did this hideous reptile haunt my dreams, that many times the very same dream was broken up in the very same way. I heard gentle voices speaking to me (I hear everything when I am sleeping), and instantly I awoke; it was broad noon, and my children were standing, hand in hand, at my bedside, come to show me their coloured shoes, or new frocks, or let me see them dressed for going out. No experience was so awful to me, and at

the same time so pathetic, as this abrupt translation from the darkness of the infinite to the gaudy summer air of highest noon, and from
the unutterable abortions of miscreated gigantic vermin, to the sight
of infancy and innocent *human* creatures.

And yet again:

Somewhere, but I knew not where—somehow, but I knew not how—
by some beings, but I knew not by whom—a battle, a strife, an agony
was travelling through all its stages—was evolving itself like the
catastrophe of some mighty drama, with which my sympathy was
the more insupportable, from deepening confusion as to its local
scene, its cause, its nature, and its undecipherable issue. I had the
power, and yet had not the power to decide it. I had the power, if I
could raise myself to will it; and yet again had not the power, for the
weight of twenty Atlantics was upon me, or the oppression of inex-
piable guilt. "Deeper than ever plummet sounded," I lay inactive.
Then, like a chorus, the passion deepened. Some greater interest
was at stake—some mightier cause than ever yet the sword had
pleaded, or trumpet had proclaimed. Then came sudden alarms,
hurryings to and fro, trepidations of innumerable fugitives, I knew
not whether from the good cause or the bad; darkness and lights;
tempest and human faces; and at last, with the sense that all was
lost, female forms, and the features that were worth all the world to
me; and but a moment allowed, and clasped hands, with heart-
breakings, partings, and then—everlasting farewells! And with a
sigh such as the caves of hell sighed when the incestuous mother
uttered the abhorred name of Death, the sound was reverberated—
everlasting farewells! And again and yet again, reverberated—
everlasting farewells!

And I awoke in struggles and cried aloud, "I will sleep no more!"

These visions, and those of a like character, in which the Malay
and the outcast girl appear and re-appear, are almost repeated again in
a work of more recent years, the production of another mind and a
widely different character. Whoever has read Kingsley's *Alton Locke*,
cannot fail to have been struck with the vivid opium-like dreams which
pass through the brain of the hero when struck down by fever. One
could almost imagine that its author had himself suffered some of the

fearful experiences which De Quincey narrates. In these the place once occupied by the two persons above named, are usurped by the cousin and Lillian; change the names, and apart from the intimate connection of the two with each other, one could almost believe himself reading a continuation of those dreams which an unfortunate accident prevented the English opium-eater giving to the world.

I was wandering along the lower ridge of the Himalaya. On my right the line of snow peaks showed like a rosy saw against the clear blue morning sky. Raspberries and cyclamens were peeping through the snow around me. As I looked down the abysses I could see far below, through the thin veils of blue mist that wandered in the glens, the silver spires of giant deodars, and huge rhododendrons, glowing like trees of flame. The longing of my life to behold that cradle of mankind was satisfied. My eyes revelled in vastness, as they swept over the broad flat jungle at the mountain foot, a desolate sheet of dark gigantic grasses, furrowed with the paths of the buffalo and rhinoceros, with barren sandy water courses, desolate pools, and here and there a single tree, stunted with malaria, shattered by mountain floods; and far beyond the vast plains of Hindostan, enlaced with myriad silver rivers and canals, tanks and rice fields, cities with their mosques and minarets, gleaming among the stately palm-groves along the boundless horizon. Above me was a Hindoo temple, cut out of the yellow sandstone. I climbed up to the higher tier of pillars among monstrous shapes of gods and fiends, that mouthed and writhed and mocked at me, struggling to free themselves from their bed of rock. The bull Nundi rose and tried to gore me; hundred-handed gods brandished quoits and sabres around my head; and Kali dropped the skull from her gore-dripping jaws to clutch me for her prey. Then my mother came, and seizing the pillars of the portico, bent them like reeds; an earthquake shook the hills—great sheets of woodland slid roaring and crashing into the valleys. A tornado swept through the temple halls, which rocked and tossed like a vessel in a storm: a crash— a cloud of yellow dust which filled the air—choked me—blinded me— burned me—

And Eleanor came by and took my soul in the palm of her hand, as the angel did Faust's, and carried it to a cavern by the sea-side and dropped it in; and I fell and fell for ages. And all the velvet mosses, rock flowers, and sparkling spars and ores, fell with me, round me, in showers of diamonds, whirlwinds of emerald and ruby, and pattered into the sea that moaned below and were quenched; and the light lessened above me to one small spark, and vanished; and I was in darkness, and turned again to my dust.

Sand—sand—nothing but sand! The air was full of sand, drifting over granite temples, and painted kings and triumphs, and the skulls of a former world, and I was an ostrich, flying madly before the simoon wind, and the giant sand pillars, which stalked across the plain hunting me down. And Lillian was an Amazon queen, beautiful, and cold, and cruel; and she rode upon a charmed horse, and carried behind her on her saddle, a spotted ounce, which was my cousin; and, when I came near her, she made him leap down and course me. And we ran for miles and for days through the interminable sand, till he sprang on me, and dragged me down. And as I lay quivering and dying, she reined in her horse above me, and looked down at me with beautiful pitiless eyes; and a wild Arab tore the plumes from my wings, and she took them and wreathed them in her golden hair. The broad and blood-red sun sank down beneath the sand, and the horse and the Amazon and the ostrich plumes shone blood-red in his lurid rays.

I was a baby ape in Borneon forests, perched among fragrant trailers and fantastic orchis flowers; and as I looked down, beneath the green roof, into the clear waters, paved with unknown water-lilies on which the sun had never shone, I saw my face reflected in the pool— a melancholy, thoughtful countenance, with large projecting brows— it might have been a negro child's. And I felt stirring in me, germs of a new and higher consciousness—yearnings of love towards the mother ape, who fed me, and carried me from tree to tree. But I grew and grew; and then the weight of my destiny fell upon me. I saw year by year my brow recede, my neck enlarge, my jaw protrude, my teeth become tusks—skinny wattles grew from my cheeks—the animal faculties in me were swallowing up the intellectual. I watched in myself, with stupid self-disgust, the fearful degradation which goes on from youth to age in all the monkey race, especially in those which approach nearest to the human form. Long melancholy mopings, fruitless strugglings to think, were periodically succeeded by wild frenzies, agonies of lust, and aimless ferocity. I flew upon my brother apes, and was driven off with wounds. I rushed howling down into the village gardens, destroying everything I met. I caught the birds and insects, and tore them to pieces with savage glee. One day, as I sat among the boughs, I saw Lillian coming along a flowery path—decked as Eve might have been the day she turned from Paradise. The skins of gorgeous birds were round her waist; her hair was wreathed with fragrant tropic flowers. On her bosom lay a baby—it was my cousin's. I knew her, and hated her. The madness came upon me. I longed to leap from the bough and tear her limb from limb; but brutal terror, the dread of man which is the doom of beasts, kept me rooted to my place. Then my cousin came, a hunter missionary; and I heard him talk to her with pride of the new world of civilisation and Christianity, which he was organising in that tropical wilderness. I listened with a dim jealous understanding—not of the words, but of the facts. I saw them instinctively, as in a dream. She pointed up to me in terror and disgust, as I sat gnashing and gibbering overhead. He threw up the muzzle of his rifle carelessly and fired—I fell dead, but conscious still. I knew that my carcase was carried to the settlement; and I watched while a smirking, chuckling, surgeon dissected me, bone by bone, nerve by nerve. And as he was fingering at my heart, and discoursing sneeringly about Van Helmont's dreams of the Archæus, and the animal spirit which dwells within the solar plexus, Eleanor glided by again like an angel, and drew my soul out of the knot of nerves, with one velvet finger tip.

Here are dreams which, however natural in their realisation to the opiophagi, are enough to cause a hearty utterance of those lines by Keats:

> *O dreams of day and night!*
> *O monstrous forms! O effigies of pain!*
> *O spectres busy in a cold, cold gloom!*
> *O land-eared Phantoms of black weeded pools!*

The "dream fugue" of the author of the *Confessions* is a day dream—a splendid one—but the type of many another dream, perhaps, that had coursed through the mind of its writer while under the influence of the subtle drug. One might almost venture the assertion that none but the "opium-eater" could have conceived and written that "fugue." But "shadows avaunt," we have stern realities yet from the pandemonium of opium.

The mind suffers and it re-acts upon the body. Although pictures of both the mental and bodily afflictions of indulgers in opium are likely to be gazed upon with somewhat of scepticism, and justly too, in these times of prejudice and outcry against opium trading, yet the stubborn fact stares the scepticism out of countenance, in many of the details of the excesses of the victims of the insinuating poppy juice. Some of these facts come to us with so high an authority and are so often repeated, that the eye and ear refuse to close and be blind and deaf to the pains which succeed the pleasures of opium.

A young eagle said to a thoughtful and very studious owl, "It is said there is a bird called Merops, which, when it rises into the air, flies with the tail first and the head looking down to the earth. Is it a fact?"

"By no means" said the owl, "it is only a silly fiction of mankind. Man himself is the Merops, for he would willingly soar to heaven, without losing sight of the world for a single instant."

Dr. Medhurst thus describes the opium-smoker of China:

> The outward appearances are sallowness of the complexion, blood-less cheeks and lips, sunken eye, with a dark circle round the eyelids, and altogether a haggard countenance. There is a peculiar appear-ance of the face of a smoker not noticed in any other condition; the skin assumes a pale waxy appearance, as if all the fat were removed from beneath the skin. The hollows of the countenance, the eyelids, fissure and corners of the lips, depression at the angle of the jaw, temples, etc., take on a peculiar dark appearance, not like that result-ing from various chronic diseases, but as if some dark matter were deposited beneath the skin. There is also a fulness and protrusion of the lips, arising perhaps from the continued use of the large mouth-piece peculiar to the opium-pipe. In fine, a confirmed opium-smoker presents a most melancholy appearance, haggard, dejected, with a lack-lustre eye, and a slovenly, weakly, and feeble gait.

Mustapha Shatoor, an opium-eater of Smyrna, took daily three drachms of crude opium. The visible effects at the time were the spar-kling eyes and great exhilaration of spirits. He found the desire of increasing his dose growing upon him. He seemed twenty years older than he really was—his complexion was very sallow, his legs small, his gums eaten away, and his teeth laid bare to the sockets. He could not rise without first swallowing half a drachm of opium. This case is detailed in the *Philosophical Transactions*, and for its veracity the philosophers are responsible.

Pouqueville says, "Always beside themselves, the Theriakis are incapable of work, they seem no more to belong to society. Toward the end of their career, they, however, experience violent pains, and are devoured by constant hunger, nor can their paregoric in any way relieve their sufferings; they become hideous to behold, deprived of their teeth, their eyes sunk in their heads, in a constant tremour, they cease to live long before they cease to exist."

Heu Naetse, a native Celestial, in his address to the Sacred Emperor, the brother of the Sun and Moon, informs his imperial majesty, that "when any one is long habituated to inhaling opium, it becomes necessary to resort to it at regular intervals, and the habit of using it, being inveterate, is destruction of time, injurious to property, and yet dear to one even as life. Of those who use it to great excess, the breath becomes feeble, the body wasted, the face sallow, and the teeth black. The individuals themselves clearly see the evil effects of it, yet cannot refrain from it. It will be found on examination that the smokers of opium are idle, lazy vagrants, having no useful purpose before them."

Dr. Ball states, "that throughout the districts of China may be seen walking skeletons—families wretched and beggared by drugged fathers and husbands—multitudes who have lost house and home dying in the streets, in the fields, on the banks of the river, without even a stranger to care for them while alive, and when dead left exposed to view till they become offensive masses."

A Pinang surgeon says that

the hospitals and poorhouses are chiefly filled with opium-smokers. In one that I had charge of, the inmates averaged sixty daily, five-sixths of whom were smokers of chandu. The effects of this habit on the human constitution are conspicuously displayed by stupor, forgetfulness, general deterioration of all the mental faculties, emaciation, debility, sallow complexion, lividness of lips and eyelids, languor and lack lustre of eye; appetite either destroyed or depraved. In the morning these creatures have a most wretched appearance, evincing no symptoms of being refreshed or invigorated by sleep, however profound. There is a remarkable dryness or burning in the throat, which urges them to repeat the opium-smoking. If the dose be not taken at the usual time, there is great prostration, vertigo, torpor, and discharge of water from the eyes. If the privation be complete, a still more formidable train of phenomena takes place—coldness is felt all over the body, with aching pains in all parts, the most horrid feelings of wretchedness come on, and if the poison be withheld, death terminates the victim's sufferings. The opium-smoker may be known by his inflamed eyes and haggard countenance, by his lank and shrivelled limbs, tottering gait, sallow visage, feeble voice, and the death boding glance of his eye. He seems the most forlorn creature that treads the earth.

The Abbè Huc writes, "nothing can stop a smoker who has made much progress in this habit, incapable of attending to any kind of business, insensible to every want, the most hideous poverty; and the sight of a family plunged into despair and misery, cannot rouse him to the smallest exertion, so complete is the disgusting apathy to which he is sunk."

The evidence of Ho King Shan is, that "it impedes the regular performance of business; those in places of trust who smoke fail to attend personally even to their most important offices. Merchants who smoke fail to keep their appointments, and all their concerns fall behind hand. For the wasting of time and the destruction of business, the pipe is unrivalled."

Oppenheim declares "that when the baneful habit has become confirmed, it is almost impossible to break it off. His torments, when deprived of the stimulant, are as dreadful as his bliss is complete when he has taken it. Night brings the torments of hell, day the bliss of paradise; and after long indulgence, he becomes subject to nervous pains, to which opium itself brings no relief. He seldom attains the age of forty, if he has begun the practice early."

Also Dr. Madden: "The debility, both moral and physical, attendant on the excitement produced by opium is terrible; the appetite is soon destroyed, every fibre in the body trembles, the nerves of the neck become affected, and the muscles get rigid. Several of these I have seen in this place at various times, who had wry necks and contracted fingers, but still they cannot abandon the custom; they are miserable until the hour arrives for taking their daily dose; and when its delightful influence begins, they are all fire and animation."

A native literati of Hong-Kong affirms, "that from the robust who smoke, flesh is gradually consumed and worn away, and their skin hangs down like bags; the faces of the weak who smoke are cadaverous and black, and their bones naked as billets of wood."

Also Dr. Oxley of Singapore: "The inordinate use of the drug most decidedly does bring on early decrepitude, destructive of certain powers connected with the increase of the species, and a morbid state of all the secretions. But I have seen a man who had used the drug for fifty years in moderation without evil effects, and one I recollect in Malacca who had so used it was upwards of eighty. Several in the habit of smoking assured me, that in moderation, it neither impaired the functions nor shortened life, at the same time they fully admitted the deleterious effects of too much."

Dr. Little visited on one occasion an opium shop, and found there two women smoking the drug—one had been a smoker for ten years.

> In the morning when she awakes she says, "I feel as one dead. I cannot do anything until the pipe is consumed. My eyelids are glazed so that they cannot be opened, my nose discharges profusely. I feel a tightness in the chest, with sense of suffocation. My bones are sore, my head aches and is giddy, and I loathe the very sight of food." Within an hour I could produce a thousand of those creatures; and if I stood at the door of an opium shop, and watched those that entered, out of the hundred would be found at least seventy-five or eighty whose appearance would not require the confession that their health was destroyed, and their mind weakened, since the day that they were cursed with the first taste of an opium-pipe. To finish this subject let me record my opinion, the result of extensive investigation. That the habitual use of opium not only renders the life of the man miserable, but is a powerful means of shortening that life.

To the last conclusion there are many objectors; and this subject has been canvassed as much as any in connection with the habit. Some years ago a trial took place in consequence of the death of the Earl of Mar, who was an opiophagi, and the insurance society on this ground objected to pay the money to his representatives. Dr. Christison, after detailing the facts, adds, "they would certainly tend on the whole rather to show that the practice of eating opium is not so injurious, and an opium-eater's life not so uninsurable, as is commonly thought." The result of the above-named trial was that the money had to be paid.

Before passing from this Plutonian region, the evidence of a good authority may be taken to show how apt prejudice is to impute even worse effects to the "subtle drug" than circumstances will warrant. An opium den is visited; the members of this convivial society are good-humoured and communicative.

> One was a chair-cooly, a second was a petty tradesman, a third was a runner in a mandarin's yanum; they were all of that class of urban population which is just above the lowest. They were, however, neither emaciated nor infirm. The chair-cooly was a sturdy fellow, well capable of taking his share in the porterage of a sixteen-stone mandarin; the runner seemed well able to run, and the tradesman, who said he was thirty-eight years old, was remembered by all of us to be a singularly young-looking man of his age. He had smoked opium for seven years. As we passed from the opium-dens, we went into a Chinese tea-garden—a dirty paved court, with some small trees and flowers in flower-pots—and a very emaciated and yawn-ing proprietor presented himself. "The man has destroyed himself by opium-smoking," said an English clergyman who accompanied us. The man being questioned, declared that he had never smoked an opium-pipe in his life—a bad shot, at which no one was more amused than the reverend gentleman who had fired it.
>
> I only take the experiment for what it is worth. There must be very many most lamentable specimens of the effects of indul-gence in this vicious practice, although we did not happen to see any of them that morning. They are not, however, so universal, nor even so common, as travellers who write in support of some thesis, or who are not above truckling to popular prejudices in England are pleased to say they are. But if our visit was a failure in one respect, it was fully instructive in another. In the first house we visited, no man spent on an average less than 80 cash a-day on his opium-pipe. One man said he spent 120. The chair-cooly spends 80, and his average earnings are 100 cash a-day. English physi-cians, unconnected with the missionary societies, have assured me that the cooly opium-smoker dies, not from opium, but from starvation. If he starves himself for his pipe, we need not ask what happens to his family. (*Times*)

CHAPTER XIII

OPIUM MORALS

Fal.   No abuse, Hal.

Poins. No abuse!

Fal.   No abuse, Ned, in the world; honest Ned, none.
       I dispraised him before the wicked, that the wicked
       might not fall in love with him; in which doing, I have
       done the part of a careful friend, and a true subject.
       No abuse, Hal; none, Ned, none; no, boys, none.

                       King Henry IV, part II

Scarce a flower that graces the earth, or a tree waving in the forests, has had its character assailed so mercilessly as the poppy. Not one of the simples or compounds of the chemist's store, even including arsenic and strychnine, has been so strictly interrogated as to the honourable and dishonourable of its intentions. It is matter of surprise that the East India Company has not been obliged, by authority of Act of Parliament, to imprint the decalogue, at least in the Chinese language, upon every cake or ball of opium leaving their stores. Take upon credit all that some men would tell you, and there would not be room for doubt, were the next informant to state that on the arrival of a cargo of opium, at such a port, on such a day, the entire population cut each other's throats, on account of the pestilential miasma diffused by the said cargo. What are really the moral effects of opium-smoking, can best be collected from a statement of facts, the reader drawing his own inferences: they are, at any rate, bad enough without the aid of exaggeration.

At Singapore stands a house of correction, in which, during the month of July, 1847, might be found forty-four Chinese criminals; and of these, thirty-five were opium-smokers—not moderate smokers, but indulgers to excess—not confining themselves to what they could obtain with such money as they could spare from their wages, but in some instances, swallowing or smoking them all up, and in certain instances, even more than their wages.* The aggregate amount of the monthly wages of seventeen of these men was £16 0s. 10d., or individually 18s. 10¹/₂d. The monthly consumption of opium of these men amounted in value to £20 16s. 3d., or individually to £1 4s. 5¹/₂d, so that each of these men, in addition to spending all his wages, begged, borrowed, or stole 5s. 7d. monthly, to make up his quantity of opium alone, without reference to any other necessaries. One of these men, who spent £1 5s. monthly, and whose wages only reached half of that amount, was asked to explain how it was to be accounted for. Was there not some error in the calculation, or was he deceiving the person to whom circumstances were being detailed? How was it possible that, with an income of only 12s. 6d., he could spend £1 5s.? The answer was a graphic one and much to the point: "What am I in here for?" Of course, the tenants of a jail can account for such discrepancies in arithmetic. The offences for which these persons were confined were such as would stand in a calendar under the rank of vagrants, suspicious characters, persons attempting to steal, and such like—the crimes committed being against *property* and not *persons*. This distinction deserves notice, as it will serve as the bases of some future suggestions.

In looking down the column of the table in which the above instances occur, it will be seen that one planter, whose income was twelve shillings and sixpence, expended in opium six times that amount; and another, whose income is not stated, but which would not far exceed the former, expended twelve times that amount in the drug. Occasional instances occur in which, where the income reached twelve shillings and sixpence, the expenditure amounted only to a trifle beyond; and where the income was sixteen shillings and eightpence, the expenditure was only eight shillings and fourpence or ten shillings.

The inspector of the above institution states:

> During the course of these investigations, I found some opium-smokers, who declared that their wages only equalled the value of the opium consumed, and in the majority of cases but little exceeded their consumption; yea, I found instances, and these not

---

*See Table XV in the Appendix.

few, where the value of the opium consumed monthly, was more than the whole wages received. The idea then suggested itself to me, that there must be an affinity betwixt opium-smoking and crime; for when once the habit is formed, it cannot be broken off, while the desire increases with the consumption. It must happen that the wages of the individual will at last be inadequate to supply his desire, even supposing that, after a lengthened career of indulgence, he was able to earn the same amount of money as when, strong, vigorous, and unimpaired, he commenced his dissipation. I, therefore, was not at all surprised when I went to the house of correction to find that three-fourths of the prisoners were opium-smokers.

An examination of the prisoners in jail in July of the same year, under different sentences, showed that out of fifty-one Chinese prisoners, fifteen only were not opium-smokers. Seventy per cent. were addicted to the vice, each consuming quantities ranging from twelve to one hundred and eighty grains per day. The same jail was again visited, and the prisoners examined a month afterwards, several fresh criminals had entered, others had been enlarged. At this time, there were sixty-nine criminals, and of these only thirty-one were opium-smokers, being only forty-five per cent. against the seventy per cent. of the former visit.

A quantity of criminals from Pinang under sentence of transportation showed, on examination, the following results: Out of twenty-one criminals, Chinese and Malays, eight did not smoke. The crimes of these men were murder, stabbing with intent to murder, burglary, and larceny. Ten of these men were Chinese, all of whom smoked but one. Of these nine, eight were condemned for offences against property, one only against the person. Of the nine persons out of the twenty-one who were convicted for offences against the person, four did not smoke, three smoked but little. Hence the conclusion is inevitable, that the criminals of the worst degree, or those committing offences against the person, are either not smokers at all, or are so only to a moderate extent. Other statistics show that, for crimes of this character, highway robbery, and burglary, forty to fifty per cent. only indulge in opium; whilst for vagrancy, misdemeanour, and petty larceny, seventy to eighty per cent. indulged in the use of the drug, and often to a very extraordinary extent.

Why do we find that those charged with the gravest offences are the least addicted to opium? May it not be that this class of criminal requires a certain ingenuity, an amount of method and calculation, and mental vigour and excitement of the passions, greater than the debased opium-smoker is possessed of, the want of which, therefore, unfits him for carrying out any such enterprises requiring such adjuncts, leaving him only capable of being a criminal on a small scale. It is well known that the Chinese are inveterate gamblers; but it is not in connexion with the pipe, but with the arrack-cup, that this vice is indulged in. The influences of opium are sedative and soothing, those of arrack stimulating and exciting; the latter, therefore, as may be supposed, is the companion of the gambler, rather than the former. There are other phases in which the two vices of opium-smoking and intoxication may be compared. The abuse of ardent spirits leads to crimes against the person; the abuse of opium leads to crimes against property. The victim of ardent spirits commits his crimes while under their influence; the devotee to opium, while under its influence, is at peace with all mankind, and dreams only of his own happiness. The drunkard, when not under the influence of liquor, may be a moral member of society, and often a contrite one; the opium-smoker at that time is often scheming the violation of moral and social laws, which, when effected, makes him a criminal, but enables him to gratify his appetite.*

De Quincey compares the two habits, not so much for the purpose of showing the tendency of either of them to crime, but for proving that opium does not produce intoxication any more than would a rump steak.

------

*Dr. Hobson states, in an official communication to the Government, "I do not know of any mortal disease from opium corresponding to *delirium tremens* from alcohol. I have never been called to attend to any accidents resulting from opium similar to those occurring so frequently from habits of intoxication from liquor. The opium-smoker, when under the full influence of his delicious drug, brawls and swaggers not in the public streets, like a drunkard, to the annoyance of bystanders, but reposes quietly on his couch, without molesting those around him."

Also Dr. Traill, of Singapore, from his own experience, has not found opium-smoking in any way so powerful a promoter of disease as the habitual use of intoxicating liquors.

The pleasure given by wine is always rapidly mounting, and tending to a crisis, after which as rapidly it declines; that from opium, when once generated, is stationary for eight or ten hours. The first— to borrow a technical distinction from medicine—is a case of acute, the second of chronic pleasure; the one is a flickering flame, the other a steady and equable glow. But the main distinction lies in this, that whereas wine disorders the mental faculties, opium, on the contrary (if taken in a proper manner) introduces amongst them the most exquisite order, legislation, and harmony. Wine robs a man of self-possession; opium sustains and reinforces it. Wine unsettles the judgment, and gives a preternatural brightness and a vivid exaltation to the contempts and the admirations, to the loves and the hatreds of the drinker; opium, on the contrary, communicates serenity and equipoise to all the faculties, active or passive, and with respect to the temper and moral feelings in general, it gives simply that sort of vital warmth which is approved by the judgment, and which would probably always accompany a bodily constitution of primeval or antediluvian health. Wine constantly leads a man to the brink of absurdity and extravagance, and beyond a certain point, it is sure to volatize and disperse the intellectual energies; whereas opium always seems to compose what had been agitated, and to concentrate what had been distracted. In short, to sum up all in one word, a man who is inebriated, or tending to inebriation is, and feels that he is in a condition which calls up into supremacy the merely human, too often the brutal part of his nature; but the opium-eater, simply as such, assuming that he is in a normal state of health, feels that the divine part of his nature is paramount, that is, the moral affections are in a state of cloudless serenity, and high over all, the great light of the majestic intellect.

It is not to be wondered at that the abuse of opium should be a fertile source of poverty, when so much of the wages of many of its votaries are devoted to it. This diseased habit is progressive, and the quantity taken must be daily increased to produce the necessary effects; but the capability of furnishing the means does not keep pace with the desire of consumption. The cooly, who, when strong and vigorous, could earn twenty-five shillings per month, has only to commence opium-smoking, and in two years he will not receive more than two-thirds of that amount, whilst he still smokes his quantity of opium; and as years roll on, he finds that, mainly on account of the vice he has adopted, he can no longer endure the toil that formerly was to him only as child's play, the amount of excitement having still to be kept up under a decreased income, he has to lessen his expenditure for clothes, and then for food, and lastly, the quantity of opium itself; until worn out, exhausted, and diseased, he finds himself the inmate of a jail or a poorhouse. A sad reflection, truly, but a history repeated over and over again, with but little variation in the lives of thousands of Chinamen and Malays.

Were poverty to be succoured in places where this description of persons most do congregate, as it is at home, thousands would become public burdens; but there the hand of charity has been closed, and the springs of compassion for the poor dried up. In Singapore, it was not until the horrid spectacle of miserable Chinese daily crawling in front of their doors, exposing their loathsome sores and leprous bodies, and polluting the air they breathed; it was not until these wretched beings, without food or friends, and deprived of the power of supporting themselves, laid them down to die in the streets, of disease and starvation, that by the active philanthropy of two or three individuals a shed was erected to keep these paupers out of sight. When the novelty passed away, the philanthropy declined, and the monthly contribution dwindled down to about three pounds, which was the sum total of the public charity of the European residents in behalf of the diseased poor of Singapore. In this shed were to be found two classes of persons, united in the same individuals, the *diseased poor*. These are the only kind of poor that excite *any* sympathy in such places, and an examination of the inmates of the *shed* will give some insight into the propensities of this class. Out of 125 under relief at the time, 70 were opium-smokers and 55 were not (or would not acknowledge it). Of these 70, some before their admission, were reduced to the alternative of *Tye* or *Samshing*, or no opium at all. The total consumption of these paupers before their admission amounted to upwards of four pounds (2022 grains) daily, giving an average daily consumption to each smoker of upwards of 28 grains, being nearly the average consumption of the opium-smoker in general, under more favourable circumstances. The greatest consumption of any one of these individuals had amounted to 120 grains, but at that rate his finances soon failed him, and he had to be content with one fourth of that amount shortly before he became an invalid. Sixty-two of these men consumed opium to the monthly value of £38 7s. 6d., while their aggregate income amounted in the same period to but £50 11s. 3d.; or, individually, the value of each man's monthly consumption of opium was 12s. 4¹/₂d., and his income was but 16s. 6d., leaving only about 4s. monthly, or 1s. per week to feed, clothe, and house himself, and in fact, for every other purpose for which money is required. Some of these did not confine themselves to this. Fifteen of them (as will be seen from Table 16) consuming all, or more than their income in opium. Surely such men were worthy not only of a pauper hospital, but also of a jail.

These paupers at one time all received even more than the average amount of wages, sufficient to have clothed and fed them and their families, and kept them comfortable, whilst at that time they were dependent on a charity which allowed them to exist on the rice which was supplied to them, and five doits a day or about a shilling per month. Thousands more, not incapacitated so much by disease as to be unable to work and not therefore inmates of the hospital, were no better off, for what they had they spent in chandu.

The Dutch Commissioners report that, "the use of opium is so much more dangerous, because a person who is once addicted to it can never leave it off. To satisfy that inclination he will sacrifice everything, his own welfare—the subsistence of his wife and children, and neglect his work. Poverty is the natural consequence, and then it becomes indifferent to him by what means he may content his insatiable desire after opium; so that at last he no longer respects either the property or life of his fellow creature."

A Chinaman, who himself is a smoker and consumes opium to the monthly value of £2, says that in one hundred Chinese about Hong-Kong and Singapore, seventy of them smoke, and that all the coolies do so more or less. If a cooly earns £1 monthly, 4s. goes for food, 10d. for house rent, a small outlay for a jacket and trowsers once in six months, and all the rest goes in opium. From his own experience, and what he has seen of others, he would say if a man had been accustomed to smoke opium for seven or eight years, and gives it up for a day he is attacked with diarrhœa, while during the time he is smoking the opposite is the case. And he who uses six grains a day will soon require twelve.

To give up opium-smoking, after it has once been commenced, all declare to be a very difficult achievement. A Malay who was apprehended on some criminal charge some years ago, when locked up, previous to examination was, as a matter of course, deprived of opium for some days, he pined away so rapidly that, although only four or five days in the lock-up house, he could not leave it when released, but was carried out, having entered the place as strong and muscular a man as can be met with.

Dr. Oxley states, "that the lower class of Chinese when deprived of their allowance, are very liable to become dropsical. The effect of deprivation at first appears to produce desperation, a heart-rendering despondency, something like the low state of delirium tremens, but differing in many respects from that malady. Death certainly does occur from deprivation, and generally by dropsy."

A great many women smoke, generally the wives of opium-smokers. A woman was discovered by a surgeon in Singapore in an opium shop up stairs smoking away, as she had done for three years, at the rate of thirty-six grains a day. She stated that she had two children, but that they were very sickly and always crying. And how did she stifle their cries? She conveyed from her lips to those of the child the fresh drawn opium vapour, which the babe inspired. This was repeated twice, when it fell back a senseless mass into its mother's arms, and allowed her quietly to finish her unholy repast. This practice she had often recourse to, as her child was very troublesome, adding that it was no uncommon thing for mothers to do.

Another inveterate opium-smoker makes his "confession," that after his quantity is consumed, he feels no desire for sleep until twelve or two in the morning, when he falls into disturbed slumbers, which last till eight or nine. When he awakes, his head is giddy, confused, and painful—his mouth is dry, he has great thirst, he has no appetite, can neither read nor write, suffers pains in all his bones and muscles, gasps for breath; he wishes to bathe, but cannot stand the shock. This state continues till he gets his morning pipe, when he can eat and drink a little, and after that attend to his business. The force of example taught him this habit, and he knows no class of people exempt from it except the Europeans. "Look," says he, appealing to himself, "I was, ere I gave way to this accursed vice, stout, strong, and able for anything. I loved my wife and children, attended to my business, and was happy; but now I am thin, meagre, and wretched. I can receive enjoyment from nothing but the pipe, my passions are gone, and if I am railed at, and abused like a dog, I return not an angry word."

Although opium-smoking is carried to such an excess among some of the Chinese coolies, yet there is no gambling amongst them at the opium shops at Singapore. It is true that this vice has been suppressed, but it is not secretly indulged in; and a gentleman who was formerly the opium farmer, says, "that the consumption of opium is but little affected by gambling, from arrack or samshu being the intoxicating medium used, a much better instrument for raising excitement and stimulating to excessive play than opium, whose effects are much more sedative than exciting."

The consideration of the morals and influence of these customs leads us to a remarkable passage in one of M. Quetelet's works, it refers to the certainty of natural laws in states as well as individuals: "All those things which appear to be left to the free will, the passions, or the degree of intelligence of men, are regulated by laws as fixed, immutable, and eternal as those which govern the phenomena of the natural world. No one knows the day or the hour of his own death; and nothing appears more entirely accidental than the birth of a boy or of a girl in any given case. But how many out of a million men living together in one country, shall have died in ten, twenty, forty, or sixty years, how many boys and girls shall be born in a million of births; all this is as certain, nay, much more certain, than any human truth."

The statistics of courts of justice have disclosed to us the regular repetition of the same crimes, and have established the fact—incomprehensive to our understandings, because we do not know the connecting links—that in every large country, the number of offences, and of each kind of offence, may be predicted for every coming year, with the same certainty as the number of the births and of the natural deaths. Of every 100 persons accused before the supreme tribunal in France, 61 are condemned; in England, 71. The variations on an average, amount hardly to $1/100$th part of the whole. We can predict with confidence, for fifteen years to come, the number of suicides generally—that of the cases of suicide by fire-arms, and that of the cases of suicide by hanging.

Every large number of phenomena of the same kind, which rise and fall periodically, leads to a fixed proportion. This is the law of large numbers to which all things and all events without exception, are subject. These laws have nothing to do with the essence of vice and virtue in the moral world, but with the external causes, and the effects they produce in human society. No one denies the influence of education, and of habits of labour and order on the conduct of men, but no one thinks of regarding this moral conduct as a mere result of those habits. Good education and improved cultivation diminish the number of offences, as well as that of the annual deaths in our tables of mortality.

The results, therefore, of a collection of statistical information carefully arranged for Singapore, one of the most inveterate of opium localities, should, on comparison with the results obtained from other quarters, show that the per centage of deaths is greater, the per centage of births less; the per centage of criminals higher, and of suicides larger, in this population of opium-smokers, than in any other equally conditioned country in which opium is indulged, or it is not proven that the habit tends to shorten life, decrease production, increase crime, and induce suicide, all of which charges have been made against it.

With this evidence we are not at present satisfactorily supplied. That opium has an influence, though probably only a minor one, on moral and social development, is not to be denied. Because man is so entirely a creature of relation, that nothing is unimportant to him. "If the movements of the remotest star that glitters in the heavens affect those of our earth, assist in determining its position in space, its climate, its productions, and thus influence the lot of man, who is the creature of these circumstances; what combinations subsisting upon the surface of the earth, or developing themselves in the bosom of society, can be deemed wholly indifferent to his conduct, and without power over his well being and happiness?"

If, as Dr. Lyon Playfair recently noticed, it is worthy of observation, that the character of the nations through which Dr. Livingstone passed in his recent travels, depended upon the habits of the people, in the acquisition of their food, as well as upon the food itself, we may expect to find opium exerting also its influence. If, for instance, the Kaffirs who lived by hunting, and were flesh-eaters, were wild and warlike; and the Wampoos, who lived principally on grain, were of a more quiet and peaceable disposition. Then again, the Bechuanos, who lived upon grain, were more civilized than the Kaffirs, and the

Macololas, who combined as their food both grain and flesh, did not
lose the warlike character, and made incursions upon their more feeble
neighbours. It was an axiom amongst the latter people, that if it were
not for the gullet (alluding to their appetites) there would be no war
or fighting amongst mankind. In those parts, such as Loando, where
the people lived upon starchy varieties of food, they had become di-
minutive in their stature; and this applied not merely to the natives,
but also to the Portuguese settlers there, for they had lost the physical
characters of their ancestors, and had become feminine in their frames
and habits, and this extended even to their handwriting. Where more
nitrogenous food was taken, the physical character of the people had
not undergone that very marked change. If food exerts this influence
upon the people of a country or district, we cannot doubt that any
habit, such as smoking tobacco or opium, chewing betel or coca, must
exert some influence upon the nations so indulging, whether that influence be good or bad.

Who will say that tobacco has no portion in the formation of the
German character? Yet the subtle and profound Germans exhibit no
extraordinary evidence in their national character of the baneful influences on their moral and social development, by their indulgence
in this habit. Compare with them the Turks and Chinese, and let the
balance be shown in favour of the most elevated in the ranks of civilization. Yet the most deficient must claim the influence of other
equally potent circumstances in extenuation, for neither opium nor
tobacco moulds the entire national character, it is only one of many
influences. Let the Papuan stand beside the Chinaman and the Turk,
and in spite of opium, the Papuan standard will exhibit a woeful short-
coming. The waters of the great Amazon river must exert some influence on the currents of the Atlantic, but none will venture to assert
that therefore the influx of such a body of water, vast in itself, but
small in comparison to the whole, is the cause of the gulf stream. The
drinking of tea will bear just such a relation to the currents in the life
of nations who indulge in that luxury, but who will declare that the
Chinese soldiers fly from the points of the British bayonets, or are
expert in the carving of ivory balls, because they indulge in a beverage
admired by other old ladies who can neither run nor carve. Neither
because certain Javanese or Malays, under the influence of an over
dose of opium, will "run amok," or other Arabs, intoxicated with

"haschish," have made the name of assassin to become an object of dread, is it to be concluded hence that all men who indulge in the use of either of these narcotics will be dangerous members of society, or that they will rush into the jaws of death without a shudder at the sight of his fangs.

Is it because the Scot loves whiskey that he is generally so cautious and shrewd in his business transactions as to win himself a name? Is it because the Cockney imbibes sundry deep potations of London porter or gin, that the enterprise and commerce of those great citizens of the world have become the envy of surrounding nations? Or is it because the Russian persisted in his love of raw turnip and sour quass, that the Malakoff and Sebastopol passed into the hands of the frog-eating Frenchman, and the beef-eating Englishman?

May we not impute to beef and tobacco, gin and opium, porter and hemp, results infinitely in advance of their power?

Dr. Eatwell writes,

> It has been too much the practice with narrators who have treated on the subject, to content themselves with drawing the sad picture of the confirmed opium debauchee, plunged in the last stage of moral and physical exhaustion, and having formed the premises of their argument of this exception, to proceed at once to involve the whole practice in one sweeping condemnation. But this is not the way in which the subject can be treated; as rational would it be to paint the horrors of *delirium tremens*, and upon that evidence, to condemn at once the entire use of alcoholic liquors. The question for determination is not what are the effects of opium used to excess, but what are its effects on the moral and physical constitution of the mass of the individuals who use it habitually, and in moderation, either as a stimulant to sustain the frame under fatigue, or as restorative and sedative after labour, bodily or mental. Having passed three years in China, I may be allowed to state the results of my observation, and I can affirm thus far, that the effects of the abuse of the drug do not come very frequently under observation; and that when cases do occur, the habit is frequently found to have been induced by the presence of some painful chronic disease, to escape from the sufferings of which the patient has fled to this resource. That this is not always the case, however, I am perfectly ready to admit, and there are, doubtless, many

who indulge in the habit to a pernicious extent, led by the same morbid impulses which induce men to become drunkards in even the most civilized countries; but these cases do not, at all events, come before the public eye. It requires no laborious search in civilized England to discover evidences of the pernicious effects of the abuse of alcoholic liquours: our open and thronged gin-palaces, and our streets, afford abundant testimony on the subject; but in China this open evidence of the evil effects of opium is at least wanting. As regards the effects of the habitual use of the drug on the mass of the people, I must affirm that no injurious results are visible. The people generally are a muscular and well-formed race, the labouring portion being capable of great and prolonged exertion under a fierce sun, in an unhealthy climate. Their disposition is cheerful and peaceable, and quarrels and brawls are rarely heard amongst even the lower orders, whilst in general intelligence, they rank deservedly high amongst orientals.

The proofs are still wanting to show that the moderate use of opium produces more pernicious effects upon the constitution, than does the moderate use of spirituous liquors, whilst at the same time, it is certain, that the consequences of the abuse of the former are less appalling in their effect upon the victim, and less disastrous to society at large, than are the consequences of the abuse of the latter. Compare the furious madman, the subject of *delirium tremens*, with the prostrate debauchee, the victim of opium; the violent drunkard, with the dreaming sensualist intoxicated with opium; the latter is at least harmless to all except to his wretched self, whilst the former is but too frequently a dangerous nuisance, and an open bad example to the community at large.

CHAPTER XIV

FALSE PROPHETS

If your wish be rest,
Lettuce and cowslip wine probatum est.

Pope

Before describing any of the imitations of opium, or substitutes for it in any form, it will not be out of place to notice briefly the tinctures in popular use in which that drug forms a prominent ingredient. *Laudanum* is the spirituous infusion, and contains the active ingredients of a twelfth part of its weight of opium. Scotch *paregoric elixir* is a solution in ammoniated spirit, and is only one-fifth of the strength of laudanum, containing, therefore, one part in sixty of opium. English *paregoric* is a tincture of opium and camphor, and is four times weaker still. The *black drop*, and *Battley's sedative liquor*, are believed to be solutions of opium in vegetable acids, and to possess, the one of them, four, and the other, three times the strength of laudanum. Although some good authorities consider this an exaggerated computation of the strength of the latter two, and that they are not more than half that strength. There are several other pharmaceutical preparations into which opium enters as a component, but to which it is unnecessary to refer. Those already named, as has before been intimated, are used not a little, to still the sounds of those miniature human organs so distasteful to bachelor ears. The practice, unfortunately so prevalent, of soothing infants with preparations of opium, cannot be too strongly deprecated. We are ready to express our surprise that oriental mothers should transfer their cigars from their own mouths to those of their infants, that the helpless little creatures may enjoy the luxury of a suck, while, at the same time, we are inuring them to the use of a far more insidious and deadly poison. Rather let us for the future, when inclined to charge this as a crime upon others, remember that scene which took place eighteen hundred years ago, and the rebuke with which it closed, in words written with the finger upon the ground, "Let him that is without sin amongst you cast the first stone at her."

One of the most important of opium substitutes is derived from a plant in itself not only harmless, but extensively used as an article of food: it is *Lactucarium* or lettuce-opium, and is prepared generally from the wild lettuce, although similar properties exist to a more limited extent in the cultivated varieties which find their way to our tables.

There is no certainty about the period at which lettuce was introduced into this country, although the time has been fixed at 1520, when it is stated to have been brought from Flanders. In the early part of the reign of Henry VIII, when Queen Katherine wished for a salad, she despatched a messenger to Holland or Flanders; at that period, therefore, very few English tables could ever boast the honour of a salad. In the privy purse expenses of Henry VIII, in 1530, an item occurs from which we learn that the gardener of York Place received a reward for bringing "lettuze" and cherries to Hampton Court. This was policy on the part of the King, his royal consort having a liking for salads, for it was rather expensive as well as tedious, to send for them to the gardens of Brabant.* In 1600, peas, beans, and lettuce were in common use in England; and in 1652, a writer of the time speaks of lettuce as a plant with which the public generally had been long familiar. One variety of the cultivated lettuce was doubtless derived from the island of Cos, inasmuch as it still bears that name.

Lettuces were known to the ancients. Dioscorides and Theophrastus

---

*Dr. Doran says that a salad was so scarce an article during the early part of the last century, that George I was obliged also to send to Holland to procure a lettuce for his queen. These vegetables must, therefore, have become unpopular before that time, or the cultivation had been for some cause discontinued, otherwise we cannot reconcile this with the fact that lettuces were common enough a century before a George sat upon the English throne.

speak of them as cultivated by the Greeks, and also used in medicine; the prickly lettuce is still found wild on the higher hills of Greece, and was probably one of the species to which the above-named ancient authors refer. Several varieties of the garden lettuce were used in salads by both Greeks and Romans. The pride of the garden of Aristoxenus was his lettuces, and he irrigated them with wine.

Two species of wild lettuce are found in Britain, the acrid and the prickly lettuce, both of which possess similar properties, yielding a juice from which lactucarium may be prepared. Two other wild species are only occasional. The lactucarium of the London Pharmacopœia is prepared only from the garden lettuce, but the acrid lettuce is stated to yield a much larger quantity and of superior quality. A single plant of the garden lettuce will yield only 17 grains of lactucarium, on an average, while a plant of the acrid lettuce yields no less than 56 grains, or more than three times that quantity; and although the milkiness of the juice increases till the very close of the time of flowering, or till the month of October in this climate, the value of the lactucarium is deteriorated after the middle of the period of flowering, for subsequently, while the juice becomes thicker, a material decrease takes place in the proportion of bitter extract contained in it.

Lactucarium is a reddish brown substance with a narcotic odour and bitter taste, having a considerable resemblance to opium. On analysis it yields a snow white crystalline substance called *lactucin*, which is narcotic in its effects. Dr. Duncan recommended the use of lactucarium as a substitute for opium, the anodyne properties of which it possesses, without being followed with the same injurious effects. In France, a water is distilled from lettuce, and used as a mild sedative. Experiments of the effects of lettuce-opium upon animals are detailed by Orfila, who states that three drachms introduced into the stomach of a dog killed it in two days, without causing any remarkable symptoms; two drachms applied to a wound in the back induced giddiness, slight sopor, and death in three days; and thirty-six grains injected, in a state of solution, into the jugular vein caused dullness, weakness, slight convulsions, and death in 18 minutes.

In North America the prickly lettuce is more common than with us, and from it the American lactucarium is extracted. In Guinea a species of lettuce is found wild, possessing precisely similar properties, and applicable to a like use. This plant is largely used by the negroes as a salad and also as an opiate.

The plants cultivated for the sake of the juice are grown in a rich soil, with a southern aspect. In such a situation they thrive vigorously, and send up thick, juicy, flower stems. As soon as these have attained a considerable height, and before the flowers expand, a portion of the top is cut off. The milky juice quickly exudes from the wound, while the heat of the sun renders it so viscid that, instead of flowing down, it concretes on the stem in a brownish flake. After it has acquired a proper consistency it is removed. As the juice closes up the vessels of the plant, another slice is taken off lower down the stem, and the juice again flows freely and another flake is formed. The same process is repeated as long as the plant affords any juice. To the crude juice, thus obtained, the name of lactucarium has been given.

"This," says Johnston,

> is one of those narcotics in which many of us unconsciously indulge. The eater of green lettuce as a salad, takes a portion of it in the juice of the leaves he swallows; and many of my readers, after this is pointed out to them, will discover that their heads are not unaffected after indulging copiously in a lettuce salad. Eaten at night, the lettuce causes sleep; eaten during the day, it soothes and calms and allays the tendency to nervous irritability. And yet the lover of lettuce would take it very much amiss if he were told that he ate his green leaves, partly at least, for the same reason as the Turk or the Chinaman takes his whiff from the tiny opium pipe: that, in short, he was little better than an opium-eater, and his purveyor than the opium smuggler on the coast of China.

Lest this should occasion some alarm in the breasts of those who prefer their lobsters with a salad, let us strive to administer a little consolation. We have seen that the cultivated or garden lettuce does not contain so much as one third the quantity of lactucarium yielded by the wild species, ten good lettuces must therefore be eaten before sufficient extract will have been consumed to have killed a dog in two days. This is upon the presumption that the lettuces eaten as salad are in precisely the same condition, and capable of affording the same amount of the extract as when cultivated specially for that purpose; but this is not the case, it is not until just before flowering that the full amount of juice is contained in the plant, a per centage only of which exists in the younger plants as gathered for the table. Nor is that quantity of the same narcotic quality as in the more matured plant, which has collected, at that period, all its strength properly to produce, and bring to perfection, its flowers and fruit.

> *Nothing hath got so far,*
> *But man hath caught and kept it as his prey.*
> *His eyes dismount the highest star,*
> *He is in little all the sphere.*
> *Herbs gladly cure our flesh, because that they*
> *Find their acquaintance there.*
>
> *More servants wait on man*
> *Than he'll take notice of: in every path*
> *He treads down that which doth befriend him,*
> *When sickness makes him pale and wan,*
> *Oh, mighty love! Man is one world, and hath*
> *Another to attend him.*

The lacticiferous or milk-bearing plants are nearly all of them connected by very important ties with man and civilization. The phenomena themselves are well worthy of study, and their association with humanity replete with interest. These plants are by no means restricted to one genus or family, nor are their properties of the same character. The one circumstance of their secreting a white juice resembling milk in appearance is almost all they have in common. In the poppy it becomes *opium*, in the lettuce *lactucarium*. It constitutes refreshing beverages, obtained in large quantities, in the sunny climes of Asia, from the cow-tree of South America, the kiriaghuma and hya-hya of British Guiana, the *Euphorbia balsamifera* of the Canary Islands, the juice of which as a sweet milk, or evaporated to a jelly, is taken as a great delicacy, and the Banyan tree, all of which, to a certain extent, supply the place of the cow, in places and conditions wherein cows are not to be found. Similar juices are collected in the form of India rubber or caoutchouc, a substance so invaluable in the arts of life. They exude from figs, euphorbiæ, and cacti, in the East Indies, South America, and Africa, from all of which places a large quantity of the consolidated juice is exported to the markets of Europe and North America. The greater quantity of these lactescent juices are elaborated in the Tropics. Gutta percha and allied substances are similarly produced, and indeed, numerous plants are possessed of this kind of secretion, which have not yet been made available for economical purposes, but which may become equally well known, and useful, to succeeding generations. Narcotic properties do not appear to be so common in these juices as the irritant or acrid, which abound in some euphorbiaceous plants, and the inert, and when coagulated and dry, elastic properties found in the siphonias, figs, and sapotaceous plants.

In St. Domingo, a species of *Muracuja* is believed to possess qualities very similar to opium, from which, and from an allied plant, Dr. Hamilton believes, that the concentrated sap, collected at a proper time, strained, evaporated, and properly prepared, would prove an excellent substitute for the expensive opium, at a cheaper rate. The species indigenous to Jamaica, is known as bull-hoof or Dutchman's laudanum. At a time when opium was scarce, from some accidental cause, in the island of Jamaica, a Dutch surgeon found in this plant a successful substitute. The plant is common in Jamaica and some other of the West Indian islands. It is an elegant climber, bearing bright scarlet blossoms, somewhat resembling a passion flower. Browne says, that the flowers are principally employed, and when infused, or mixed in a state of powder with wine or spirits, are regarded as a safe and effectual narcotic.

Dr. Landerer states that the Syrian rue is a highly esteemed plant in Greece. This plant appears to have been known to the ancients, and mentioned by Dioscorides. Its properties are narcotic, resembling those of the Indian hemp. The Turks macerate the seeds in scherbet or boosa, administering the infusion internally. It also serves in the preparation of a yellow dye. The seeds are sometimes used by the Turks as a spice, and the same people also resort to them to produce a species of intoxication. The Emperor Solyman, it is stated, kept himself in a state of intoxication by their use. The peculiar phenomena of this intoxication has not, that we are aware, been described, but we are informed that the property of producing it exists in the husks of seeds, from which a chemical principle of a narcotic nature has been obtained.

There is another plant, a native of Arabia, and of the nightshade family, so prolific in narcotics, the seeds of which are used by some of the Asiatics to produce those mental reveries and excitement so much

coveted. These seeds, the produce of a plant known to botanists under the name of *Scopolia mutica*, are also roasted and infused to form a
sort of drink, in which the Arabs and some others indulge.

The seeds of a species of *Sterculia* are said to be used by the natives
of Silhet as a substitute for opium. The Cola nuts, so highly esteemed
by the negroes of Guinea, are the produce of a Sterculia. The natives
attribute very extraordinary properties to these seeds, somewhat analogous to those claimed by the Peruvians for the leaf of the coca, stat-
ing, that if chewed, they satisfy hunger, and prevent the natural crav-
ing for food, that for this purpose they carry some with them when
undertaking a long journey. They are also affirmed to improve the
flavour of anything that may be subsequently eaten, if a portion of
one of them is taken before meals. Formerly they were even more
esteemed than at the present day. In those times, fifty of them were
sufficient to purchase a wife. These seeds are flat, and of a brownish
colour and bitter taste. Their tonic properties have been supposed
equal to those of the famed Cedron seeds of Guiana and the more
famous Cinchona bark of the Andes. Probably further and more elaborate investigation will prove that these wonderful seeds possess slightly
beneficial properties as a tonic, it may be even inferior to those of the
roots of the Gentian, or other parts of some of our indigenous plants.

In the Straits, the leaves of the "Beah" tree are used by the opium-
smokers as a substitute for opium, when that drug is not procurable.
These serrated leaves, the produce of we know not precisely what tree,
except under the above native name, are occasionally sold in the bazaars or markets at a quarter of a rupee per catty, or at the rate,
Anglicised, of fourpence halfpenny per pound.

In addition to the substances which do duty for opium knowingly
and wittingly, there are others which enter into its composition in the
form of adulteration, to which writers on materia medica have drawn
attention, and ultimately Dr. Hassell. These also deserve, with far greater
appropriateness, the designation of false prophets, since, promising the
glimpses of paradise which opium is believed to give, they only

> *Keep the promise to the lip*
> *and break it with the heart.*

The first sophistication, says Pereira, which opium receives, is that
practiced by the peasants who collect it, and who lightly scrape the

epidermis from the shells or capsules to augment the weight. This operation adds about one-twelfth of foreign matters, which are removed by the Chinese in their method of preparing the opium and forming it into chandu.

According to Dr. Eatwell, the grosser impurities usually mixed with the drug to increase its weight are mud, sand, powdered charcoal, soot, cow dung, pounded poppy petals, and pounded seeds of various descriptions. All these substances are readily discoverable in breaking up the drug in cold water, decanting the lighter portion, and examining the sediment. Flour is a very favourite article of adulteration, but is readily detected. Opium so adulterated becomes sour, breaks with a short ragged fracture, the edges of which are dull, and not pink and translucent as they should be. The farina of the boiled potato is not unfrequently made use of; ghee and ghour (an impure treacle) are occasionally used, as being articles at the command of most of the cultivators. Their presence is revealed by the peculiar odour and consistence which they impart to the drug. In addition to the above, a variety of vegetable juices, extracts, pulps, and colouring matters are occasionally fraudulently mixed with the opium, such as the inspissated juice of the prickly pear, the extracts prepared from the tobacco plant, the thorn apple, and the Indian hemp. The gummy exudations from various plants are frequently used; and of pulps, the most commonly employed are those of the tamarind, and of the Bael fruit. To impart colour to the drug various substances are employed, as catechu, turmeric, the powdered flowers of the mowha tree, etc. Here is a list long enough to satisfy any antiquarian, containing delicacies of all kinds, the essence of which would improve any soothing syrup or Godfrey's cordial, with which, under the name of opium, they may be incorporated, whether they may consist of tobacco juice, cow dung, or bad treacle.

Let us still enlarge the collection from the experience of Dr. Normandy, eminent in chemical analysis—

Opium is often met with in commerce from which the morphine has been extracted; on the other hand, this valuable drug is often found adulterated with starch, water, Spanish liquorice, lactucarium, extract of poppy leaves, of the sea-side poppy, and other vegetable extracts, mucilage or gum tragacanth, or other gums, clay, sand, gravel. Often the opium is mixed in Asia and Egypt, when fresh and soft, with finely bruised grapes, from which the stones have been removed; sometimes also a mixture, fabricated by bruising the exterior skins of the capsules and stalks of the poppy together with the white of eggs, in a stone mortar, is added in certain proportions to the opium. In fact, this most valuable drug, certainly one of the most important, and most frequently used in medicine, is also one of the most extensively adulterated.

Dr. Landerer has described an adulteration of a sample of opium obtained direct from Smyrna; it consisted of salep powder in large proportions, and he was afterwards informed that this is a very common adulteration, practised in order to make the opium harder, and to hasten the process of drying. Dr. Pereira speaks of an opium which contained a gelatiniform substance, and Mr. Morson met with opium in which a similar substance was present. Dr. Landerer also states that the extract obtained by boiling the poppy plants is commonly added to Smyrna opium.

Dr. Hassell found "that out of twenty-three samples of opium analysed, nineteen were adulterated, and four only genuine, many of these as shown by the microscope, being adulterated to a large extent; the prevailing adulterations being with poppy capsules and wheat flour," in addition to which adulteration two samples of Smyrna opium, and two of Egyptian opium were adulterated with sand, sugar, and gum.

From the analysis of forty samples of powdered opium, he found also, "that thirty-three of the samples were adulterated, and one only genuine; the principal adulterations, as in the previous case, being with poppy capsules and wheat flour. That four of the samples were further adulterated by the addition of powdered wood, introduced, no doubt, in the process of grinding."

Dr. Thomson stated in his evidence before the Parliamentary Committee, that he had known extract of opium mixed with extract of senna, and from thirty to sixty per cent. of water.

Dr. O'Shaughnessy found from 25 to 21 per cent. of water in Indian opium (Behar agency), and 13 per cent. in Patna opium.

Dr. Eatwell, the opium examiner in the Benares district, finds that the proportion of water varies from 30 to 24–25 per cent. in the opium of that district.

In 1838, a specimen of opium resembling that of Smyrna, was presented to the Societe de Pharmacie of Paris, being part of a considerable quantity which had been introduced into commerce at Paris and Havre. It did not exhibit the least trace of morphia. It was in rolls, well covered with leaves, had a blackish section, and a slightly elastic consistence. It became milky upon contact with water. Its odour and taste were analogous to opium, but feebler. It was adulterated with so much skill, that agglutinated tears appeared even under a magnifier—a character which had hitherto been regarded as decisive in detecting pure opium, but which with this occurrence lost its value. The same article appears to have been met with also in the United States.

A writer in Singapore states,

I lately saw a Chinaman brought to the police for fabricating opium balls. The imitation balls were composed of a skin or husk formed from the leaves of Madras tobacco, inside was sand, which was evidently intended to form the shape of the balls till the outer covering had sufficiently set, the whole was neatly sewed with bandages of calico, which would be removed when the tobacco was able to retain its proper shape, the sand would then be abstracted, and a mixture of gambier and opium substituted, while the outside would be rubbed over with a watery solution of chandu. By these means the native traders are much and often imposed upon.

CHAPTER XV
NEPENTHES

Bright Helen mixed a mirth-inspiring bowl,
Tempered with drugs of sovereign use, to assuage
The boiling bosom of tumultuous rage;
To clear the cloudy front of wrinkled care,
And dry the tearful sluices of despair.

Pope's *Homer*

The influence of climate in modifying the characters of plants is a circumstance known to all botanical students. The same plant, in temperate regions and under the tropics, exhibits different properties, or, we should rather say, in one instance developes more highly certain properties which in the other lie nearly dormant. The newly-introduced sorghum, from which we have been promised an unfailing supply of excellent sugar, fails in the North of France to reach that degree of maturity, or to develope in such manner its saccharine secretions as to be available for the manufacture of a crystallizable sugar. The sweet floating grass (*Glyceria fluitans*) in Poland and Russia supplies farinaceous seeds, which, under the name of manna croup, are consumed as food; but no seeds at all available for that purpose are produced at home from the same plant, although it grows freely. The flavour of the onion, as grown in Egypt, is, we are assured, far milder, and vastly different from the bulbs cultivated in Britain. The odour of violets and other flowers grown for perfumery and other purposes at Nice, have a scent more rich and delicious than when grown in English soil, subject to our variable climate. But the most extraordinary effect of all, produced by these influences upon plants, occurs in the case of hemp, which in Europe develops its fibrous

qualities to such an extent as to produce a material for cordage hitherto unsurpassed; but in India, while deficient in this respect, develops
narcotic secretions to such an extent as to occupy a prominent position among the chief narcotics of the world.

It was for some time supposed that the Indian or narcotic hemp
was a different species to that which is cultivated for textile purposes;
and even now it is often characterised by a different specific name,
which would seem to assume that the species are distinct. This, however, the most celebrated of our botanists deny. The difference is declared to be, not one of species, but of climate, and of climate only.
The native home of the hemp plant is assigned by Dr. Lindley to Persia and the hills in the North of India, whence it has been introduced
into other countries. Burnett says, "Hemp seed is nutritious and not
narcotic; it has the very singular property of changing the plumage of
bullfinches and goldfinches from red and yellow to black, if they are
fed on it for too long a time or in too large a quantity." Never having
tried the experiment, we have no ground for disputing or authority for
verifying these remarks. If such, however, is the case, hemp seed possesses some property, if not narcotic, which canary and poppy seeds,
we should presume, do not.

Johnny Englishman, with his usual genius for discovery and invention, has been discovered filling his pipe on board ship with oakum, when the stores of tobacco have been exhausted, but not being satisfied from his own experiments of the superiority of hemp, in
that form, to his brother Jonathan's tobacco, he therefore adheres to
the latter. He considers hemp an excellent thing when twisted into
a good hawser, but does not like it as "twist" in the masticatory ac-
ceptation of the term; nor does he at all admire the twist of Ben
Battle, when

> Round his melancholy neck
> A rope he did entwine,
> And for his second time in life,
> Enlisted in the line.
>
> One end he tied around a beam,
> And then removed his pegs;
> And as his legs were off, of course
> He soon was off his legs.

And there he hung till he was dead
As any nail in town;
For though distress had cut him up,
It could not cut him down.

Hemp is one of those plants which adapts itself well to any climate: there is scarce a country in Europe where it cannot, or might not, be cultivated. From Poland and Russia in the North, to Italy in the South, the fibre is supplied to our markets. In North America it is grown for its fibre, and in South America it is grown for its narcotic properties. Throughout Africa, it may be found chiefly as an article for the pipe. In most of Asia it is known, and it has been cultivated in Australia. Thus, in its distribution, it may now be considered as almost universal.

Twenty-five centuries ago, Herodotus wrote of its cultivation by the Scythians: "They have a sort of hemp growing in this country very like flax, except in thickness and height; in this respect the hemp is far superior—it grows both spontaneously and from cultivation, and from it the Thracians make garments very like linen, nor would any one who is not well skilled in such matters distinguish whether they are made of flax or hemp; but a person who has never seen this hemp, would think the garment was made of flax." Then follows a description of the use of the hemp as a narcotic: "The Scythians, transported with the vapour, shout aloud." Antiquity is in favour of this narcotic, and its use for that purpose before any other, except perhaps the poppy, was known, or at least of those now in use. The *nepenthes* of Homer has been supposed to have been this plant, or one of its products. The use of hemp had become so general amongst the Romans at the time of Pliny, that they commonly made ropes and cordage of it. The practice of chewing the leaves to produce intoxication existed in India in very early ages, whence it was carried to Persia, and before the middle of the thirteenth century, this custom was adopted in Egypt, but chiefly by persons of the lower orders.

The narcotic properties of hemp become concentrated in resinous juice, which in certain seasons and in tropical countries exudes, and concretes on the leaves, slender stems, and flowers. This constitutes the base of all the hemp preparations, to which all the powers of the drug are attributable. In Central India, the hemp resin called *chur-*

rus, is collected during the hot season in the following manner. Men clad in leathern dresses run through the hemp fields, brushing through the plants with all possible violence; the soft resin adheres to the leather, and is subsequently scraped off and kneaded into balls, which sell at from five to six rupees the seer, or about five or six shillings per pound. A still finer kind, the *momeca* or waxen churrus, is collected by the hand in Nepaul, and sells for nearly double the price of the ordinary kind. Dr. McKinnon says—"In Nepaul, the leathern attire is dispensed with, and the resin is collected on the skin of naked coolies." In Persia the churrus is obtained by pressing the resinous plant on coarse cloths, and then scraping it from these and melting it in a pot with a little warm water. Mirza considers the churrus of Herat the most powerful of all the varieties of the drug. The hemp resin, when pure, is of a blackish grey colour, with a fragrant narcotic odour, and a slightly warm, bitterish, acrid taste.

The dried hemp plant which has flowered, and from which the resin has been removed, is called in India *gunjeh*. It sells at from twelve annas to a rupee the seer, or from ninepence to a shilling per pound, in the Calcutta bazaars. It is sold chiefly for smoking, in bundles two feet long and three inches in diameter, containing twenty-four plants. The colour is dusky green, the odour agreeably narcotic, the whole plant resinous and adhesive to the touch.

The larger leaves and capsules without the stalks, are called *bang*, *subjee*, or *sidhee* in India, and have been brought into the London market under the name of *guaza*. They are used for making an intoxicating drink, for smoking, and in the conserve called *majoon*. Bang is cheaper than Gunjeh, and though less powerful, is sold at so low a price that for one halfpenny enough can be purchased to intoxicate an habituated person.

The Gunjeh consumed in Bengal comes chiefly from Mirzapore and Ghazeepur, being extensively cultivated near Gwalior and in Tirhoot. The natives cut the plant when in flower, allow it to dry for three days, and then lay it in bundles averaging two pounds each which are distributed to the licensed dealers. The best kinds are brought from Gwalior and Bhurtpore, and it is cultivated of good quality in gardens around Calcutta.

The *majoon* or hemp confection, is a compound of sugar, butter, flour, milk, and bang. The mass is divided into small lozenge-shaped pieces; one dram will intoxicate a beginner, three drams one experienced in its use. The taste is sweet and odour agreeable. Most carnivorous animals will eat it greedily, and very soon become ludicrously drunk, but seldom suffering any worse consequences.

The confection called *el mogen* in use amongst the Moors appears to be similar to, if not identical with, the *majoon* of India.

The ancient Saracens and modern Arabs in some parts of Turkey and generally throughout Syria, use preparations of hemp still known by the name of *haschisch* or *hashash*. M. Adolph Stuze, the court apothecary at Bucharest, thus describes the haschisch, by which general name all intoxicating drugs whose chief constituent is hemp, are well known all over the East. The tops and all the tender part of the hemp plant are collected after flowering, dried and kept for use. There are several methods of using it.

1. Boiled in fat, butter, or oil, with a little water; the filtered product is employed in all kinds of pastry.
2. Powdered for smoking. Five or ten grains of the powder are smoked from a common pipe with ordinary tobacco, probably the leaf of a species of Lobelia (Tombuki) possessing strong narcotic properties.
3. Formed with tragacanth mucilage into pastiles, which are placed upon a pipe and smoked in similar doses.
4. Made into an electuary with dates or figs and honey. This preparation is of a dark brown or almost black colour.
5. Another electuary is prepared of the same ingredients, with the addition of spices, cloves, cinnamon, pepper, amber, and musk. This preparation is used as an aphrodisiac.

The confection most in use among the Arabs is called *Dawamese*. This is mingled with other stimulating substances, so as to administer to the sensual gratifications, which appear to be the *summum bonum* of oriental existence.

The *haschisch* extract is about the consistence of syrup, and is of a dark greenish colour, with a narcotic odour, and a bitter, unpleasant taste.

A famous heretical sect among the Mahometans bore the name of Assassins, and settled in Persia in 1090. In Syria they possessed a large tract of land among the mountains of Lebanon. They assassinated Lewis of Bavaria in 1213, were conquered by the Tartars in 1257, and extirpated in 1272. Their chief assumed the title of "Ancient of the Mountain." These men, some authorities inform us, were called *Haschischins* because the use of the haschish was common among them in the performance of certain rites, and that the ancient form has been corrupted into that now in use. M. de Sacy states that the word "assassin" has been derived from the Arabic name of hemp. It has also been declared, that during the wars of the Crusades, certain of the Saracen army while in a state of intoxication from the use of the drug, rushed madly into the Christian camp, committing great havoc, without themselves having any fear of death, and that these men were called *Hashasheens*, whence has arose our word "assassin." The term "hashash" says Mr. Lane, signifies "a smoker or an eater of hemp," and is an appellation of obloquy; noisy and riotous people are often called "hashasheen," which is the plural of that appellation, and the origin of our word "assassin."

Benjamin of Tudela says, "In the vicinity of Lebanon reside the people called Assassins, who do not believe in the tenets of Mahommedanism, but in those of one whom they consider like unto the Prophet Kharmath. They fulfil whatever he commands them, whether it be a matter of life or death. He goes by the name of Sheikh-al-Hashishin, or, their old man, by whose command all the acts of these mountaineers are regulated. The Assassins are faithful to one another, by the command of their old man, and make themselves the dread of every one, because their devotion leads them gladly to risk their lives, and to kill even kings, when commanded."

In the centre of the Persian, as well as the Assyrian territory of the Assassins, that is to say, both at Alamut and Massiat, were situated, in a space surrounded by walls, splendid gardens—true eastern paradises—there were flower-beds, and thickets of fruit trees, intersected by canals; shady walks and verdant glades, where the sparkling stream bubbles at every step; bowers of roses and vineyards; luxurious halls, and porcelain kiosks, adorned with Persian carpets and Grecian stuffs, where drinking vessels of gold, silver, and crystal glittered on trays of the same costly materials; charming maidens and handsome boys, black-eyed and seductive as the houris and boys of Mahomed's paradise, soft as the cushions on which they reposed, and intoxicating as the wine which they presented; the music of the harp was mingled with the songs of birds, and the melodius tones of the songstress harmonised with the murmur of the brooks—everything breathed pleasure, rapture, and sensuality.

A youth who was deemed worthy, by his strength and resolution, to be initiated into the Assyrian service, was invited to the table and conversation of the grand master or grand prior; he was then intoxicated with henbane (haschish) and carried into the garden, which, on awakening, he believed to be paradise. Everything around him, the houris in particular, contributed to confirm his delusion. After he had experienced as much of the pleasures of paradise—which the prophet has promised to the blessed—as his strength would admit, after quaffing enervating delight from the eyes of the houris and intoxicating wine from the glittering goblets, he sank into the lethargy produced by debility and the opiate, on awakening from which, after a few hours, he again found himself by the side of his superior. The latter endeavoured to convince him that corporeally he had not left his side, but that spiritually he had been wrapped into paradise, and had then enjoyed a foretaste of the bliss which awaits the faithful, who devote their lives to the service of the faith and the obedience of their chief. Thus did these infatuated youths blindly dedicate themselves as the tools of murder, and eagerly sought an opportunity to sacrifice their terrestrial, in order to become the partakers of eternal life.

To this day, Constantinople and Cairo show what an incredible charm opium with henbane exerts on the drowsy indolence of the Turk and the fiery imagination of the Arab, and explains the fury with which those youths the enjoyment of these rich pastiles (haschish), and the confidence produced in them, that they are able to undertake anything or everything. From the use of these pastiles they were called Hashishin (herb-eaters), which, in the mouths of Greeks and Crusaders, has been transformed into the word Assassin, and as synonymous with murder, has immortalized the history of the order in all the languages of Europe.*

This is the account given by Marco Polo, as repeated by Von Hammer in his *History of the Assassins*. To this let us further add M. Sylvestre

---

*Von Hammer's *History of the Assassins*.

de Sacy's, from a memoir read before the Institute of France:

> I have no doubt whatever, that denomination was given to the Ismaelites, on account of their using an intoxicating liquid or preparation, still known in the East by the name of haschish. Hemp leaves, and some other parts of the same vegetable, form the basis of this preparation, which is employed in different ways, either in liquid or in the form of pastiles, mixed with saccharine substances, or even in fumigation. The intoxication produced by the haschish, causes an ecstasy similar to that which the orientals produce by the use of opium; and from the testimony of a great number of travellers, we may affirm that those who fall into this state of delirium, imagine they enjoy the ordinary objects of their desires, and taste felicity at a cheap rate. It has not been forgotten that when the French army was in Egypt the General-in-chief Napoleon, was obliged to prohibit, under the severest penalties, the sale and use of these pernicious substances, the habit of which has made an imperious want in the inhabitants of Egypt, particularly the lower orders. Those who indulge in this custom are to this day called Hashishin, and these two different expressions explain why the Ismaelites were called by the historians of the Crusades sometimes Assissini and sometimes Assassini.

As an instance of the blind submission of these devoted followers to the will of their chief, it is narrated that Jelaleddin Melekshah, Sultan of the Seljuks, having sent an ambassador to the Sheikh of the Assassins, to require his obedience and fealty, the son of Sahab called into his presence several of the initiated. Beckoning to one of them, he said, "Kill thyself," and instantly stabbed himself; to another, "Throw thyself down from the rampart"; the next instant he lay a mutilated corpse in the moat. On this the grand master, turning to the envoy, who was unnerved by terror, said— "In this way am I obeyed by seventy thousand faithful subjects. Be that my answer to thy master."

From comparison of these notes, it will therefore appear that the order of Hashishans used the haschish, as a means whereby to induce young men to devote themselves to their cause. That it was used by the chief for its intoxicating and illusionary properties, probably without the knowledge of the members of the order, but as a secret, the divulging of which would have defeated his design, and that it was

not indulged in habitually by the order; but that from its use in these
initiatory rites they came to be called Haschishans, afterwards corrupted into Assassins. And ultimately, that their murderous acts procured for all those who in future times imitated them, the honour of
their name.

But to return from this long digression, we still meet with the name
of Haschisch and Hashasheen in Egypt, and also with preparations of
hemp, which are believed as of old to transport those who indulge therein
to scenes such as paradise alone is supposed to furnish.

> Where'er his eye could reach,
> Fair structures, rainbow-hued, arose;
> And rich pavilions through the opening woods
> Gleamed from their waving curtains sunny gold;
> And winding through the verdant vale,
> Flowed streams of liquid light,
> And fluted cypresses reared up
> Their living obelisks,
> And broad-leaved plane trees in long colonnades,
> O'er arched delightful walks,
> Where round their trunks the thousand-tendril'd vine
> Wound up, and hung the boughs with greener wreaths,
> And cluster not their own.

M. Rouyer, of the Egyptian Commission, says, with the leaves and
tops, collected before ripening, the Egyptians prepare a conserve, which
serves as the base of the *berch*, the *diasmouk*, and the *bernaouy*. Hemp
leaves reduced to powder and incorporated with honey or stirred with
water, constitute the *berch* of the poorer classes.

Dr. Livingstone found hemp in use among the natives of Southern Africa under the name of *mutokuane*.

With the Hottentots it is known as *Dacha*, and another plant used
for similar purposes among them is called the *wild Dagga* or Dacha.
The use of hemp as a narcotic appears to be very general in all parts of
Africa.

The D'amba possesses numerous native titles, but is only understood by those distinctive terms which the negroes give it in their
respective countries. By the people of Ambriz and Musula it is pronounced as D'yambah, while to the various races in Kaffraria, it is

more generally known under the Hottentot name of Dakka or Dacha. This plant is extensively cultivated by the Dongós, Damarás, and other tribes to the southward of Benguela. Among the Ambundas or aborigines of Angola, the dried plant is duly appreciated, not only for its narcotic effects, but likewise on account of some medicinal virtues which it has been reputed to enjoy. The markets of St. Paul de Loanda are mostly supplied from the Dongós, and other adjacent tribes, and from St. Salvador, and the towns in the vicinity of Upper Kongo.

The mode in which it is prepared for sale, consists in carefully separating from the leaves and seeds, the larger stalks, retaining only the smaller stems, which are compressed into a conical mass, varying from two to four inches in diameter, and from one to two feet in length, the whole being covered by some dried vegetable, firmly secured by thin withes. The substance thus manufactured is ordinarily employed for the purpose of smoking, and is endowed with powerful stimulant and intoxicating principles, consequently it is proportionately prized by those nations who are familiar with those peculiar qualities, and is probably viewed more in the light of a luxury owing to the absence of all other sources of excitement, for which, perhaps, it was the only available substitute.

The Zulu Kaffirs and Delagoans of the South Eastern Coast use it under the same or like names. Amongst the former the herb is powdered and used as snuff. The true tobacco is known amongst them, and is grown to a certain extent, but the use of hemp both for smoking and snuffing, is far more common. Perhaps, requiring less cultivation, it suits best their indolent habits.

The most eminent of the Persian and Arabian authors refer the origin of hemp intoxication to the natives of Hindostan. But few traces, however, of its early use can be found in any part of India.

In the *Rajniguntu*, a treatise on materia medica, the date of which is vaguely estimated at about six hundred years ago, there is a clear account of this drug. The names under which it is there known are, *bijoya*, *ujoya*, and *joya*, meaning promoters of success; *brijputta*, or the strengthener; *chapola*, the causer of a reeling gait; *ununda*, or the laughter-moving; *hursini*, the exciter of sexual desire.

In another treatise in Sanscrit, of later date, the above is repeated; and in a religious treatise, called the Hindu Tantra, it is stated that *Sidhee* is more intoxicating than wine.

In the fifth chapter of the Institutes of Menu, Brahmins are prohibited to use pabandoo or onions, *gunjara* or *gunjah*, and such condi-
ments as have strong and pungent scents.

Persian and Arabic writers give, however, a fuller and more particular account of the early use of this substance. Makrisi treats of the
hemp in his description of the ancient pleasure-grounds in the vicinity of Cairo. This quarter, after many vicissitudes, is now a mass of
ruins. In it was situated a cultivated valley, named Djoneina, which
was the theatre of all conceivable abominations. It was famous, above
all, for the sale of the *hasheesha* or haschisch, which is still consumed
by certain of the populace, and from the consumption of which sprung
those excesses which gave rise to the name of "assassin," in the time
of the Crusades. This author states that the oldest work in which hemp
is noticed is a treatise by Hassan, who states that in the year of the
Hegira 658, the Sheikh Djafar Shirazi, a monk of the order of Haider,
learned from his master, the history of the discovery of hemp. Haider,
the chief of ascetics and self-chasteners, lived in rigid privation on a
mountain between Nishabor and Rama, where he established a mon-
astery of Fakirs. Ten years he had spent in this retreat, without leaving
it for a moment, till one burning summer's day, when he departed
alone to the fields. On his return, an air of joy and gaiety was imprinted on his countenance; he received the visits of his brethern,
and encouraged their conversation. On being questioned, he stated
that, struck by the aspect of a plant which danced in the heat as if
with joy, while the rest of the vegetable creation was torpid, he had
gathered and eaten of its leaves. He led his companions to the spot—
all ate, and all were similarly excited. A tincture of the hemp-leaf in
wine or spirits, seems to have been the favourite formula in which the
Sheikh Haider indulged himself. An Arab poet sings of Haider's emerald cup—an evident allusion to the rich green colour of the tincture
of the drug. The Sheikh survived the discovery ten years, and subsisted chiefly on this herb, and on his death his disciples, by his desire,
planted an arbour in which it grew about his tomb. From this saintly
sepulchre, the knowledge of the effects of hemp is stated to have spread
into Khorasan. In Chaldea it was unknown until the Mahommedan
year 728, during the reign of the Caliph Mostansir Billah. The kings
of Ormus and Bahrein then introduced it into Chaldea, Syria, Egypt,
and Turkey.

In Khorasan, it seems that the date of the use of hemp is considered, notwithstanding the foregoing, to be far prior to Haider's era.
Biraslan, an Indian pilgrim, contemporary with Cosroes (whoever this
same Cosroes may be, for it is a name often occuring, and applied as
Cæsar or Czar to more than one generation), is stated to have introduced and diffused the custom through Khorasan and Yemen.

In 780 M.E. very severe ordinances were passed in Egypt against this
practice of indulging in hemp. The Djoneina garden was rooted up,
and all those convicted of the use of the drug were subjected to the
extraction of their teeth. But in 792 M.E. the custom re-established
itself with more than original vigour. A vivid picture is given by Makrisi
of the vice and its victims: "As a general consequence, great corruption
of sentiments and manners ensued, modesty disappeared, every base
and evil passion was openly indulged in, and nobility of external form
alone remained to those infatuated beings." In Chapter II, some further memoranda will be found of the early history of this extraordinary
narcotic.

Not only was its intoxicating power, but many other properties—
some true, some fabulous—were known at the above periods. The
contrary qualities of the plant—its stimulating and sedative effects—
are dwelt on: "They at first exhilarate the spirits, cause cheefulness,
give colour to the complexion, bring on intoxication, excite the imagination into the most rapturous ideas, produce thirst, increase appetite, excite concupiscence; afterwards, the sedative effects begin to
preside, the spirits sink, the vision darkens and weakens, and madness, melancholy, fearfulness, dropsy, and such like distempers are the
sequel." Mirza Abdul Russac says of it: "It produces a ravenous appetite and constipation, arrests the secretions, except that of the liver,
excites wild imagining, a sensation of ascending, forgetfulness of all
that happens during its use, and such mental exaltation that the be-
holders attribute it to supernatural inspiration." To which he also adds:
"The inexperienced, on first taking it, are often senseless for a day,
some go mad, others are known to die."

Whether for the purpose of increasing its power, or for what other
reason we know not, in India the seeds of Datura are mixed with hemp,
in compounding some of the confections, as well as the powder of *nux
vomica*. This is, however, exceptional, neither of these substances entering into the composition of the Majoon of Bengal any more than

does corrosive sublimate form a proportion of the pills in general use by the opium-eater of Constantinople.

It is a custom with some people to blame, without limit, those who indulge in nervous stimulants of a nature differing from their own, while serving the same purpose. Thus, one who thinks that Providence never designed his corporeal frame to become a perambulating beer-barrel, eschews all alcoholic drinks, but at the same time eschews not the abuse of those who think fit to indulge in a little wine for their stomach's sake, or a draught of porter for their bodily infirmities. These same abstainers still adhere to their tea and coffee, and though harmless enough as these dietetics may be, yet they in part serve the purposes for which others employ alcoholic stimulants. An eminent chemist states that persons accustomed to the use of wine, when they take cod liver oil, soon lose the taste and inclination for wine. The Temperance Societies should therefore canonise cod liver oil.

It is true that thousands have lived without a knowledge of tea or coffee; and daily experience teaches, that under certain circumstances they may be dispensed with without disadvantage to the merely animal vital functions. "But it is an error," writes Liebig,

> certainly, to conclude from this that they may be altogether dispensed with in reference to their effects; and it is a question whether, if we had no tea and no coffee, the popular instinct would not seek for and discover the means of replacing them. Science, which accuses us of so much in these respects, will have, in the first place, to ascertain whether it depends on the sensual and sinful inclinations merely, that every people of the globe has appropriated some such means of acting on the nervous life—from the shore of the Pacific, where the Indian retires from life for days, in order to enjoy the bliss of intoxication with coca, to the Artic regions, where the Kamtschatdale and Koriakes prepare an intoxicating beverage from a poisonous mushroom. We think it, on the contrary, highly probable, not to say certain, that the instinct of man, feeling certain blanks, certain wants of the intensified life of our times, which cannot be satisfied or filled up by mere quantity, has discovered, in these products of vegetable life, the true means of giving to his food the desired and necessary quality. Every substance, in so far as it has

a share in the vital processes, acts in a certain way on our nervous system, on the sensual appetites, and the will of man.

So, although some have no tobacco, they find in the use of hemp or opium a substitute for that vegetable which nature has denied them. There can be no doubt that had we never become acquainted with tobacco or gin, we should have discovered and used some other narcotic in the place of the one, and a no less fiery and injurious form of alcohol instead of the other. To talk of the *degraded* Chinese as *barbarians*, indulging to an awful extent in opium, and the *ignorant* Hindoo and Arab, as in madness revelling in debauches of hemp confections, is an evidence of the workings of the same narrow-minded prejudices under which some who abstain from alcoholic stimulants rail and rave at those whose feelings and habits lay in an opposite direction, charging upon the enjoyments of the many the excesses of the few. Friend Brooklove, drink thy tea, and re-consider thy verdict!

CHAPTER XVI

GUNJA AT HOME

Oh, kind and blissful mockery, when the manacled felon, on his bed of straw, is transported to the home of his innocent boyhood, and the pining and forsaken fair, is happy with her fond and faithful lover—and the poor man hath abundance—and the dying man is in joyous health—and despair hath hope—and those that want are as though they wanted not—and they who weep are as though they wept not.—But the fashion of these things passeth away.

"At home" may mean, that quarter-day has passed with all its terrors, accounts settled, bills filed, tax-collectors satisfied, and the horizon of finance clear and cloudless. There is no fear of duns or doctors, and John Thomas announces "at home." Or it may mean, that having enrobed oneself in morning gown and slippers, filled and lighted our pipe, seated ourselves in an easy chair, placed our feet firmly and contentedly on the hearthrug, and commenced enveloping ourselves in a cloud like that in which Juno conveyed the vanquished Paris from the field to the presence of the fairest of the daughters of Greece, we *feel*, with reference to ourselves, and in despite of the rest of the world—"at home." Or it may mean, that having made the "grand tour," crossed the desert on a camel, or seen the lions of Singapore, Hong-Kong, and Shanghai, we are once more on our native soil, and no longer fear Italian banditti or Turkish plague, sandstorms or crocodiles, Chinese poisoners or bow-wow pie, that we breathe again, and are "at home." Or it may mean half-a-dozen things beside. But to see a man at home, is to see him in all the gradations of light and shade, of sunlight and shadow, brighter and deeper, than when he covers his head and walks abroad to look at the sun.

Gunja is not at home in Europe. Notwithstanding the efforts made in England and France to introduce the Indian hemp into medical practice, and the asseverations of medical practitioners in British India, who have extolled its power as a narcotic and anodyne, it has never settled upon European soil. The drugs already in use to produce sleep and alleviate pain, still occupy their old popularity, undisturbed by the visit of a stranger, who, finding the reception too cold, has retreated. In France, certain experiments were made, and by leave of Dr. Moreau, we shall take advantage of them, and of the *Journal of Psychological Medicine*, to ascertain the effects of this drug on those who have used it.

Since the days of Prosper Albinus, both learned and unlearned have listened with wonder to the marvellous effects of those "drowsy syrups of the East," when—

> *Quitting earth's dull sphere, the soul exulting soars*
> *To each bright realm by fancy conjured up,*
> *And clothed in hues of beauty, there to mix*
> *With laughing spirits on the moonlight green;*
> *Or rove with angels through the courts of heaven,*
> *And catch the music flowing from their tongues.*

In Asia Minor an extract from the Indian hemp has been from time immemorial swallowed with the greatest avidity, as the means of producing the most ecstatic delight, and affording a gratification even of a higher character than that which is known there to follow on the use of opium. A small dose seems only to influence the moral faculties, giving to the intellectual powers greater vivacity, and momentary vigour. A larger dose seems to awaken a new sensibility, and call into action dormant capabilities of enjoyment. Not only is the imagination excited, but an intensity of energy pervades all the passions and affections of the mind. Memory not only recurs with facility to the past, but incorporates delusions with it, for with whatever accuracy the facts may be remembered, they are painted with glowing colours, and made sources of pleasure. The senses become instruments also of deception, the eye and the ear, not only are alive to every impression, but they delude the reason, and disturb the brain, by the delusions to which they become subject. Gaiety, or a soothing melancholy, may be produced, as pleasant or disagreeable sights or sounds are presented.

So much alive are the swallowers of haschisch to the effect of external objects upon the perceptive powers, that they generally retire to the depths of the harem, where the almas, or females educated for this purpose, add, by the charms of music and the dance, to the false perceptions which the disordered condition of the brain gives rise to. Insensibly the reason and the volition are entirely overcome, and yield themselves up to the fantastic imagery which affords such delight. Can we wonder at such people producing and admiring all the extravagancies of the *Arabian Nights' Entertainments?* Can we be surprised at their belief in a paradise for the future, which is at best but a voluptuary's dream?

At the commencement of the intoxication produced by the hemp, there is the most perfect consciousness of the state of the disordered faculties. There exists the power of analyzing the sensations, but the mind seems unwilling to resume its guiding and controlling power. It is conscious that all is but a dream, and yet feels a delight in perfect abandonment to the false enjoyment. It will not attempt to awaken from the reverie, but rather to indulge in it, to the utmost extent of which it is capable. There seems an ideal existence, but it is too pleasurable to shake off—it penetrates into the inmost recesses of the body—it envelopes it. The dreams and phantoms of the imagination appear part of the living being; and yet, during all this, there remains the internal conviction that the real world is abandoned, for a fictitious and imaginative existence, which has charms too delightful to resist. To the extreme rapidity with which ideas, sensations, desires, rush across the brain, may be attributed the singular retardation of time, which appears to be lengthened out to eternity. Similar effects, proceeding, doubtless, from the same or similar causes, are noticed in the *Confessions of an Opium-Eater,* wherein he speaks of minutes becoming as ages.

Dr. Moreau gives singular illustrations of this peculiar state. On one occasion he took a dose of the haschisch previously to his going to the opera, and fancied that he was upwards of three hours finding his way through the passage leading to it. M. de Saulcey partook of a dose of haschisch, and when he recovered, it appeared to him that he had been under its influence for a hundred years at least.

Whilst an indescribable sensation of happiness takes possession of the individual, and the joy and exultation are felt to be almost too much to be borne, the mind seems totally at a loss to account for it, or to explain from what particular source it springs. There is a positive sensation of universal contentment, but it is vain to attempt to explain the nature of the enjoyment. The peculiar motion appears to be wholly inexplicable. A sense of something unusual pervades every fibre, but all attempts to analyze or describe it are declared to be in vain. After a certain period of time the system appears to be no longer capable of further happiness, the sensibility seems thoroughly exhausted, a gentle sense of lassitude, physical and moral, gradually succeeds—an apathy, a carelessness, an absolute calm, from which no exterior object can arouse the torpid frame. These are the great characteristics of this stage. The most alarming or afflicting intelligence is listened to without exciting any emotion. The mind is thoroughly absorbed, the perception seems blunted, the senses scarcely convey any impression to the brain. A re-action has taken place, yet the collapse is unattended with any disagreeable feeling. The energies are all prostrate, yet there are none of those depressing symptoms which attend the last stages of ordinary intoxication. All that is described is an ineffable tranquillity of soul, during which it is perfectly inaccessible to sorrow or pain. "The haschisch eater is happy," continues Dr. Moreau, "not like the gourmand, or the famished man when satisfying his appetite, or the voluptuary in the gratification of his amative desires; but like him who hears tidings which fill him with joy, or like the miser counting his treasures, the gambler who is successful at play, or the ambitious man who is intoxicated with success."

All those who have tried the experiment do not speak in such glowing terms of the results. M. de Saulcey, who tried it at Jerusalem, says:

> The experiment, to which we had recourse for passing our time, turned out so utterly disagreeable that I may safely say, not one of us will ever be tempted to try it again. The haschisch is an abominable poison which the dregs of the population alone drink and smoke in the East, and which we were silly enough to take, in too large a dose, on the eve of New Year's-day. We fancied we were going to have an evening of enjoyment, but we nearly died through our imprudence. As I had taken a larger dose of this pernicious drug than my companions, I remained almost insensible for more than twenty-four hours, after which I found myself completely

broken down with nervous spasms, and incoherent dreams.

It is not uncommon for illusions and hallucinations to occur during the early stage, when the senses have lost their power of communicating faithfully to the brain the impressions they receive.

Dr. Auber, in his work on the plague, narrates various instances of
delusions occurring in the course of his administering hemp preparations as a relief in that disease. An officer in the navy saw puppets
dancing on the roof of his cabin—another believed that he transformed into the piston of a steam-engine—a young artist imagined
that his body was endowed with such elasticity as to enable him to
enter into a bottle, and remain there at his ease. Other writers speak
of individuals similarly affected: one of a man who believed himself
changed entirely into brittle glass, and in constant fear of being cracked
or broken, or having a finger or toe knocked off; another, of a youth
who believed himself growing and expanding to such an extent, that
he deemed it inevitable that the room in which he was would be too
small to contain him, and that he must, during the expansion, force
up the ceiling into the room above. Dr. Moreau, on one occasion,
believed that he was melting away by the heat of the sun, at another,
that his whole body was inflated like a balloon, that he was enabled to
elevate himself, and vanish in the air. The ideas that generally presented themselves to him of these illusions were, that objects wore
the semblance of phantasmagoric figures, small at first, then gradually
enlarging, then suddenly becoming enormous and vanishing. Sometimes these figures were subjects of alarm to him. A little hideous
dwarf, clothed in the dress of the thirteenth century, haunted him for
some time. Aware of the delusion, he entreated that the object which
kept up the illusion should be removed—these were a hat and a coat
upon a neighbouring table. An old servant of seventy-one, was, upon
another occasion, represented by his eye to the brain as a young lady,
adorned with all the grace of beauty, and his white hair and wrinkles
transformed into irresistible attractions. A friend who presented him
with a glass of lemonade was pictured to his disordered imagination as
a furnace of hot charcoal. Sometimes the happiness was interrupted
by delusions that affrighted him. Thus, having indulged himself with
his accustomed dose, every object awoke his terror and alarm, which
neither the conviction of his own mind nor the soothing explana-

tions of his friends could diminish, and he was of a considerable length of time under the most fearful impressions.

Through the darkness spread
Around, the gaping earth then vomited
Legions of foul and ghastly shapes, which
Hung upon his flight.

These are the immediate effects produced by this most extraordinary substance. There are others, however, still more singular, which have attracted the attention of travellers, and become the objects of intense curiosity. These are of a nature unknown in connection with any other substance, and have formed the basis of numerous marvellous narrations, that have astonished even the incredulous. Those who have seen the fearful symptoms betrayed during delirium tremens, and have heard the sufferers declare that they saw before them genii, fairies, devils, know how the senses may become the source of delusion, and hence may judge to *what* a disordered state of the intellect may lead. When the brain has once become disordered by the use of the narcotic hemp, it becomes ever afterwards liable to hallucinations and delusions, unlike those produced by anything else, save intoxicating liquours after an attack of delirium tremens. The mind then believes that it sees visions, and beholds beings with whom it can converse. The phenomena gradually develop themselves, until illusions take the place of realities, and hold firm possession of the mind, which would seem on all other points to be healthy and vigorous, but on this point, insane. So firm and so fixed becomes the belief, that neither argument convinces, nor ridicule shakes, the individual from his faith, in which a prejudiced or too credulous nature confirms him but the more.

The Arabs, especially those of Egypt, are exceedingly superstitious, and there is scarce a person, even among the better informed, who does not believe in the existence of genii. According to their belief there are three species of intelligent beings, namely, angels, who were created of light, genii, who were created of fire, and men, created of earth. The prevailing opinion is that Sheytans (devils) are rebellious genii. It is said that God created the genii two thousand years before Adam, and that there are believers and infidels among them as among men. It is held that they are aerial animals with transparent bodies, which can assume any form. That they are subject to death, but live many ages. The following are traditions of the prophet concerning them. The genii are of various shapes, having the forms of serpents, scorpions, lions, wolves, jackals, etc. They are of three kinds, one on the land, one in the sea, one in the air. They consist of forty troops, each troop consisting of six hundred thousand. They are of three sorts, one has wings and fly; another, are snakes and dogs; and the third move about from place to place like men. Domestic snakes on the same authority, are asserted to be genii. If serpents or scorpions intrude themselves upon the faithful at prayers, the prophet orders that they be killed, but on other occasions, first to admonish them to depart, and then if they remained to kill them. It is related that Aisheeh, the prophet's wife, having killed a serpent in her chamber, was alarmed by a dream, and fearing that it might have been a Muslim jinnee, as it did not enter her chamber when she was undressed, gave in alms, as an expiation, about three hundred pounds, the price of the blood of a Muslim. The genii appear to mankind most commonly in the shapes of serpents, dogs, cats, or human beings. In the last case, they are sometimes of the stature of men, and sometimes of a size enormously gigantic. If good, they are generally resplendently hand-some, if evil, horribly hideous. They became invisible at pleasure (by a rapid extension or rarefaction of the particles which compose them) or suddenly disappear in the earth, or air, or through a solid wall.

The Sheykh Khaleel El Medabighee related the following anec-dote of a jinnee. He had, he said, a favourite black cat, which always slept at the foot of his mosquito curtain. Once, at midnight, he heard a knocking at the door of his house, and his cat went and opened the hanging shutter of the window, and called, "Who is there?" A voice replied, "I am such a one" (mentioning a strange name), "the jinnee, open the door." "The lock," said the Sheykh's cat, "has had the name pronounced upon it." It is the custom to say, "In the name of God, the compassionate, the merciful," on locking the door, covering bread, laying down their clothes at night, and on other occasions, and this they believe protects their property from the genii. "Then throw me down," said the voice, "two cakes of bread." "The bread-basket," an-swered the cat at the window, "has had the name pronounced upon it." "Well, " said the stranger, "at least give me a draught of water." But he was answered that the water-jar had been secured in the same manner, and asked what he was to do, seeing that he was likely to die of hunger and thirst. The Sheykh's cat told him to go to the door of the next house, and went there also himself, and opened the door, and soon after returned. Next morning the Sheykh deviated from a habit which he had constantly observed; he gave to the cat half of the fateereh upon which he was wont to give, and afterwards said, "O my cat, thou knowest that I am a poor man; bring me then a little gold," upon which words the cat immediately disappeared, and he saw it no more. Such are the stories which they believe and narrate of these genii; and there is scarce an indulger in haschisch whose imagination does not lead him to believe that he has seen or had communication with some of these beings.

Mr. Lane, translator of the *Arabian Nights*, had once a humourous cook addicted to the intoxicating haschisch, of whom he relates the following circumstance:

> Soon after he had entered my service, I heard him, one evening, muttering, and exclaiming on the stairs as if surprised at some event, and then politely saying, "but why are you sitting here in the draught? Do me the favour to come up into the kitchen, and amuse me with your conversation a little!" The civil address not being answered, was repeated, and varied several times, till I called out to the man, and asked him to whom he was speaking. "The efreet of a Turkish soldier," he replied, "is sitting on the stairs, smoking his pipe, and refuses to move; he came up from the well below; pray step and see him." On my going to the stairs, and telling the servant that I could see nothing, he only remarked that it was because I had a clear conscience. My cook professed to see this efreet frequently after.

Dr. Moreau enumerates many instances, from his own immediate followers, of genii seers among the haschisch eaters. His dragoman, who had been attached in that capacity to Champollion, the captain of the vessel, and several sailors, had not only a firm belief in, but had actually received visits from genii or efreets, and neither argument nor ridicule could shake their conviction. The captain had, on two occasions, seen a jinnee, he appeared to him under the form of a sheep. On returning one evening somewhat late to his house, the captain found a stray sheep bleating with unusual noise. He took him home, sheared him for his long fleece, and was about to kill him, when suddenly the sheep rose up to the height of twenty feet, in the form of a black man, and in a voice of thunder, announced himself as a jinnee.

One of the sailors, Mansour, a man who had made nearly twenty voyages with the Europeans, recounted his interview with a genius under the guise of a young girl of eight or ten years of age. He met her in the evening on the banks of the Nile, weeping deplorably because she had lost her way. Mansour, touched with compassion, took her home with him. In the morning he mounted her on an ass, to take her to her parents. On entering a grove of palms, he heard behind him some fearful sighs; on looking round to ascertain the cause, he saw, to his horror, that the little girl has dismounted, that her lower extremities had become of an enormous length, resembling two frightful serpents, which she trailed after her in the sand. Her arms became lengthened out, her face mounted up into the skies, black as charcoal, her immense mouth, armed with crocodile's teeth, vomited forth flame. Poor Mansour fell suddenly upon the earth, where, overcome with terror, he passed the night. In the morning he crawled home, and two months of illness attested the fact of disorder of the brain.

Many such tales are recounted, and all told by the sufferers with the firmest belief, and the most earnest conviction of their truth; each, by his own delusion, strengthening and confirming others. All those who had seen visions had their minds diseased through the use of haschisch, while those who did not indulge in the habit were free from these extraordinary illusions. These hallucinations seem to be manifested independently of any then existing affection of the brain, and the individual appears, under other circumstances, fitted for the usual avocations of life. They may be only symptoms of a previously disordered intellect, but they may also be the starting point, from which insanity is developed. In all instances in which these hallucinations occur, watchfulness is necessary, since, in the majority of cases they terminate finally in derangement of the brain to the extent generally denominated *madness*.

Other curious results from the use of this narcotic are detailed by Dr. O'Shaughnessy, as exhibited by patients in India, to whom he had prescribed it, in his capacity of medical practitioner, and other experiments he made.

A dog, to whom some *churrus* was given, in half an hour became stupid and sleepy, dozing at intervals, starting up, wagging his tail as if extremely contented; he ate food greedily, on being called he staggered to and fro, and his countenance assumed the appearance of utter and helpless drunkenness. In six hours these symptoms had passed away, and he was perfectly well and lively.

A patient to whom hemp had been administered, on a sudden uttered a loud peal of laughter, and exclaimed, that four spirits were springing with his bed into the air. Attempts to pacify him were in vain, his laughter became momentarily more and more uncontrollable. In a short time he exhibited symptoms of that peculiar nervous condition, which mesmerists have of late years made us more acquainted with, under the name of catalepsy. In whatever imaginable attitude his arms and legs were placed, they became rigid and remained. A waxen figure could not be more pliant or stationary in each position, no matter how contrary to the natural influence of gravity on the part. A strong stimulant drink was given to him, and his intoxication led to such noisy exclamations, that he had to be removed to a separate room, where he soon became tranquil, in less than an hour his limbs had gained their natural condition, and in two hours he said he was perfectly well, and very hungry.

A rheumatic cooly was subjected to the influence of half a grain of hemp resin. In two hours the old gentleman became talkative and musical, told several stories, and sang songs to a circle of highly delighted auditors, ate the dinners of two persons, subscribed for him in the ward, and finally fell soundly asleep, and so continued until the following morning. At noon he was perfectly free from headache, or any unpleasant sequel; at his request, the medicine was repeated, and he was indulged with it for a few days, and then discharged.

A medical pupil took about a quarter of a grain of the resin in the form of tincture. A shout of loud and prolonged laughter ushered in the symptoms, and a state of catalepsy occurred for two or three minutes. He then enacted the part of a Rajah giving orders to his courtiers; he could recognize none of his fellow students or acquaintances—all to his mind seemed as altered as his own condition; he spoke of many years having passed since his student's days, described his teachers and friends with a piquancy which a dramatist would envy, detailed the adventures of an imaginary series of years, his travels, his attainment of wealth and power. He entered on discussions on religious, scientific, and political subjects, with astonishing eloquence, and disclosed an extent of knowledge, reading, and a ready apposite wit, which those who knew him best were altogether unprepared for. For three hours and upwards he maintained the character he at first assumed, and with a degree of ease and dignity perfectly becoming his high situation. This scene terminated nearly as abruptly as it commenced, and no headache, sickness, or other unpleasant symptoms followed the excess.

Without detailing instances in which its virtues as a medicinal agent are set forth, or naming cases of hydrophobia in which it was given and failed, or of tetanus in which it was resorted to with success, we can scarce forbear noticing the fact, that to an infant only 60 days old, 130 drops of the tincture had to be given to produce narcotism, while 10 drops produced those effects in the student above named, who believed himself an important Rajah.

The most recent information we have of the effects of haschisch is supplied by Professor K. D. Schroff. It relates to a kind called "birmingi," the laughter producer ("macht keif") obtained from Bucharest.

This preparation was in the form of tablets, hard and difficult to break, externally almost black and smooth, with but a slight smell. The taste was neither bitter nor aromatic, but rather insipid. On prolonged mastication, the very tough mass became gradually pappy, and eventually dissolved in the saliva, leaving a crumbling solid substance. It produced irritation in the throat, when chewed for a long time.

Dr. Heinrich took ten grains of this preparation in May, 1859, at about half-past five in the afternoon. He chewed this quantity for about an hour, during which it gradually dissolved and was swallowed; only the insoluble residue, about two grains, was spit out. Irritation of the throat, and slight nausea, succeeded. The attempt to smoke a cigar in the open air had to be given up on account of dryness and roughness in the throat. Dr. Heinrich walked into town, and looked at the printshops without perceiving any change in himself. At the end of an hour and a half, about seven o'clock, he met an acquaintance, to whom he talked all kinds of nonsensical trash, and made the most of foolish comparisons; henceforth, everything he looked at seemed to him ridiculous. This condition of excitement lasted about twenty minutes, during which his face and eyes grew redder and redder. Suddenly a great degree of sadness came over him everything was too narrow for him—he acquired a disturbed appearance, and became pale. His sadness increased to a feeling of anxiety, accompanied by the sensation as if his blood was flowing in a boiling state up to his head; the feeling as if his body was raised aloft, and as if he was about to fly up, was particularly characteristic. His anxiety and weakness overcame him to such a degree, that he was obliged to collect all the power of his will, and his companion had to seize him firmly under the arm, in order to bring him on, which was done in all haste, as he feared a new attack, and wished, if possible, to reach a place where he could be taken care of; but in the course of three minutes, while he was still walking, the attack set in with increased violence.

It was only with great difficulty he reached the Institute—here he immediately drank two pints of cold water, and washed his head, neck, and arms with fresh water, on which he became somewhat better. The improvement, however, lasted only about five minutes. He sat down on a chair and felt his pulse, which he found to be very small and slow, with very long intervals. He was no longer in a state to take out his watch to ascertain more exactly the frequency of his pulse, for the feeling of anxiety came over him again, and with it he traced the premonitory symptoms of a new and violent attack. He was taken into the adjoining chamber, stripped himself partly of his clothes, and gave over his things, directing what was to be done with them after his death, for he was firmly convinced that his last hour had struck, and continually cried out, "I am dying; I shall soon be undergoing dissection in the dead-room." The new attack was more violent than the former were, so that the patient retained only an imperfect degree of consciousness, and at the height of the paroxysm, even this disappeared. After the fit, too, consciousness returned but imperfectly: only so much remained in his recollection, that the images which arose within him constantly increased in ghastliness, until they gave way to the unconscious state, and that gradually, with returning consciousness, less formidable figures appeared in their stead. Subsequently he stated that it appeared to him as if he were transported from the level of surface to a hill, thence to a steep precipice, thence to a bare rock, and lastly to the ridge of a hill, with an immense abyss before him. From this time, he could no longer control the current of ideas following one another with impetuous haste, and he could not avoid speaking uninterruptedly until a fresh attack came on, which quite deprived him of consciousness for some minutes. The flow of his ideas had now free course; and notwithstanding his loquacity, he could only utter a few words of what he imagined. All his thoughts and deeds from his childhood came into his mind. The senses of sight and hearing were unimpaired, for when he opened his eyes, he knew all who were standing about him, and recognized them by their voices when his eyes were closed. Towards ten o'clock—that is, four hours and a half after the seizure—the storm was somewhat allayed; he obtained control over his imagination, ceased to speak incessantly, and traced where he felt pain. During the night he drank a great deal of lemonade; nevertheless, sleep fled from him, and his imagination was constantly at work. Next morning he dressed, and was conveyed home, but could not set to his daily work, because, notwithstanding the greatest efforts, he could not collect his scattered thoughts, and he also felt bodily weak. He was obliged to take to bed, where he remained till the morning of the third day. During this time, he drank four pints of lemonade, and took soup only twice, as he had no appetite. On the third day he was led about, supported by a second person, but was still rather confused and giddy. This day he ate but little, and drank lemonade. During the second and third nights, his sleep was tranquil. On the fourth day he felt well again, regained his appetite, his strength increased, and his appearance became less unsettled. Nevertheless, walking about for half an hour tired him much. The depression which came on after the excitement gave way only gradually.*

The incautious use of hemp is also noticed as leading to, or ending in, insanity, especially among young persons, who try it for the first time. This state may be recognised by the strange balancing gait of the victim, a constant rubbing of the hands, perpetual giggling, and a propensity to caress and chafe the feet of all bystanders, of whatever rank. The eye wears an expression of cunning and merriment which can scarcely be mistaken. In a few cases, the patients are violent—in all, voraciously hungry.

Under the influence of this drug, its devotees exhibited, doubtless, to the astonished gaze of the early travellers from this, and other northern countries, strange freaks and antics, which filled them with wonder, and sent them home brim-full of wonderful legends and mar-

---

* *Dublin Quarterly Journal of Medical Science.*

vellous stories gathered from the lips of the votaries of hemp. The
ready and active brain of the oriental—always associating places and
people, actions and accidents, men and manners, with the unseen
agency of ghosts and genii—under the influence of haschisch, gave
full scope to their imaginations, letting loose upon the traveller a tor-
rent of romance, and peopling every corner of his route with legions
of spirits, set him wondering to himself whether he had really escaped
from the common-place world of his nativity into another sphere specially devoted to the occupation of etherial beings. Now listening to
the narrative of a reputed communicant with spirits, he hears of the
concentrated genii, confined in the narrow form of a little dog, or
smaller still, in a little fish, gradually expanding, and towering higher
and higher, till his head reached to the clouds, and then with a voice
of thunder communicating his message to the terrified and superstitious Arab crouching at his feet. Anon, he hears of the plague, and his
credulous dragoman informs him that once upon a time a pious Moslem was worshipping at sunrise, when he saw a hideous phantom approaching him, and the following conversation passed between them.

"Who art thou?"

"The Plague."

"Whither goest thou?"

"To Cairo."

"Wherefore?"

"To kill ten thousand."

"Go not."

"It is destined that I should."

"Go then, but slay not more than thou hast said."

"To hear is to obey."

After the plague was over, at the same hour, and in the same place,
the phantom once more appears to him, and the holy man again addressed him thus—

"Whence comest thou?"

"From Cairo."

"How many persons hast thou destroyed?"

"Ten thousand, according to my orders."

"Thou liest, twenty thousand are dead."

"Tis true, I killed ten thousand, *fear* carried off the remainder."

Shortly, and the traveller passes a tree, a mound, or a mass of

ruins. The dragoman narrates the story of confined treasures and pro-
tecting genii, and marvels of the days long gone, and of deeds of sin,
and ends with the universal ejaculation, "God is great, and Mahomet
is his prophet." From these people of mysteries and land of marvels
the traveller returns, and though he only narrates, for fear of shame,
the more credible of the stories he has heard, from that day forth, poor
man, his friends shake their heads, and mutter their fears that a tropical sun has addled his brains.

Naturally and nationally superstitious and credulous, the use of
the narcotic assists in adding to his store of legendary lore, and the
Arab or Turk becomes in himself not only a new edition of the *Arabian Night's Entertainments*, but *it* also becomes in him a living belief,
and the narration comes from his lips with all the earnestness of positive truth, impressing itself upon the auditor as a circumstance in which
the narrator was a principal actor. And father to son, and generation
to generation, tell the tales, recount the marvels, and swallow the
haschisch of their fore-fathers, and Allah is praised, and Mahomet is
still "the Prophet."

CHAPTER XVII

ⱧUBBⱢE-BUBBⱢE

This is a strange repose, to be asleep
With eyes wide open, standing, speaking, moving,
And yet so fast asleep.

The Tempest

The *Hubble-Bubble proper* is a smoking appa-
ratus so contrived that the smoke, in its passage from the point of
consumption to that of inhalation, shall pass through water, which
performs the office of a cooler. The *Hubble-Bubble common* consists of
a cocoa-nut shell, with two holes perforated in one end, at about an
inch apart, through the germinating eyes of the nut. Through these
orifices the kernel is extracted, and a wooden or bamboo tube, about
nine inches long, surmounted by a bowl, is passed in at one opening
to the bottom of the shell, which is partly filled with water, and the
smoke is either sucked from the other hole, or a tube is inserted into
that opening also, as an improvement on the ruder practice, through
which to imbibe the smoke. The hubble-bubble is used generally for
smoking hemp, but in Siam occasionally for opium.

Smoking the hemp is indulged in, with some variations, from the
course usually pursued with tobacco. In Africa this mode of indulgence seems to be more universal than that of the Indian weed. The
inhabitants of Ambriz seek with avidity the solace of this preparation;
they, nevertheless, appear to employ it in moderation, and are not so
passionately addicted to its influence as other native tribes—they
therefore suffer less from those pernicious effects which result from

intemperate indulgence in it. The Aboriginal method of smoking this
narcotic consists in fixing the clay bowl of a native pipe into the centre of a large gourd, and passing it to each individual composing the
community, who in succession take several inhalations of the smoke,
which is succeeded by violent paroxysms of coughing, flushed face,
suffused eyes, and spasmodic gestures, with other symptoms indica-
tive of its dominant action on the system. Upon the subsidence of this
excitement, the party experience all those soothing sensations of ease
and comfort, with that pleasing languor stated to constitute the potent charm, that renders it in such universal request. If the inhaling
process is carried beyond this stage, inebriation shortly supervenes.*

ABORIGINAL DAKKA PIPE OF AMBRIZ

The Hottentots and Bushmen smoke the leaves of this plant, either alone or mixed with tobacco; and as they generally indulge to
excess, invariably become intoxicated. When the Bushmen were in
London exhibiting themselves, they smoked the hemp, from pipes
made from the tusks of animals.

The Bechuanas have a curious method of smoking the *dacha*. Two
holes the size of the bowl of a tobacco-pipe are made in the ground
about a foot apart; between these a small stick is placed, and clay
moulded over it, the stick is then withdrawn, leaving a passage connecting the two holes, into one of which the requisite material and a
light is introduced, and the smoking commenced by the members of
the party, each in turn lying on his face on the ground, inhaling a
deep whiff, and then drinking some water, apparently to drive the

---

* Dr. Daniell in *Pharmaceutical Journal.*

fumes downward. It is a singular circumstance, that a similar method of smoking is employed by certain of the tribes of India, as already described, on the authority of Dr. Forbes Royle.

EGOODU, OR SMOKING HORN, OF THE ZOOLUS

Among the Zoolus the *dacha* is placed at the end of a reed introduced into the side of an ox-horn, which is filled with water, and the mouth applied to the upper part of the horn. The quantity of smoke which is inhaled through so large an opening, unconfined by a mouthpiece, often affects the breath, and produces much coughing, notwithstanding which the natives are very fond of it; this kind of pipe is called *Egoodu*. Tobacco composed of the dried leaf of the wild hemp is in general use, and has a very stupifying effect, frequently intoxicating, on which occasions they invariably commence long and loudly to praise the king.

Though some of the Zoolus indulge in smoking, all, without exception, are passionately fond of snuff, which is composed of dried *dacca* leaves mixed with burnt aloes, and powdered. No greater compliment can be offered than to share the contents of a snuff calabash with your neighbour. The snuff is shovelled into the palm of the hand, with a small ivory spoon, whence it is carefully sniffed up. Worse than a Goth would that barbarian be who would wantonly interrupt a social party thus engaged.

The Delagoans of the eastern coast, consider the smoking of the "hubble-bubble" one of the greatest luxuries of life. A long hollow reed or cane, with the lower end immersed in a horn of water, and the upper end capped with a piece of earthenware, shaped like a thimble, is held in the hand. They cover the top, with the exception of a small aperture, through which, by a peculiar action of the mouth, they draw the smoke from the pipe above by the water below; they fill the mouth, and after having kept it some time there, eject it with violence from the ears and nostrils. "I have often," says Mr. Owen,

> known them giddy, and apparently half stifled from indulging in this fascinating luxury—it produces a violent whooping and cough-ing, accompanied by a profuse perspiration, and great temporary debility, and yet it is considered by the natives highly strengthen-ing, and is always resorted to by them previously to undertaking a long journey, or commencing work in the field. To the hut of an old man who was thus indulging himself, I was attracted by the loud-ness of the cough it had occasioned, and as I entered observed that his feeble frame had almost fallen a victim to the violent effects of the bang or dakka he was smoking. He had thrown himself back on some faggots, and it was not until I had been some time there that he appeared at all conscious of my presence; yet, as soon as the half inebriated wretch had obtained sufficient strength, he commenced his devotions to the pipe again, and by the time I quitted the hut was reduced to the same state as that in which I had found him.

"I have seen the opium-eaters of Constantinople," writes the *Times'* correspondent, "and the hashish-smokers of Constantine. I recollected having a taboosh in the bazaars of Smyrna from a young Moslem whose palsied hand and dotard head could not count the coins I offered him. I recollect the hashish-smokers of Constantine, who were to be seen and heard every afternoon at the bottom of the abyss which yawns under the Adultress Rock—lean, fleshless Arabs—smoking their little pipes of hemp-seed, chaunting and swaying their skeleton forms to and fro, shrieking to the wild echoes of the chasm, then sinking ex-hausted under the huge cactus—sights and sounds of saturnalia in purgatory."

Hemp, of all narcotics, appears to be the most uncertain in its effects. It is so in the form of haschisch or alcoholic infusion, and doubtless is so also when smoked. Professor Schroff says of it—"I have seen patients take from one to ten, or, in one case, even so much as thirty grains of the alcoholic extract in the course of an evening and night, sometimes within a few hours, without producing any particular symptoms, except some determination to the head; even the so much wished for sleep, on account of which the remedy was taken, was not obtained, while in other cases, one grain of the same preparation, from the same source, produced violent symptoms, bordering on poisoning—delirium, very rapid pulse, extreme restlessness, and subsequently, considerable depression. I must, therefore, repeat, that Indian hemp, and all its preparations, exhibits the greatest variety in the degree and mode of action, according to the difference of individuality, both in the healthy and diseased condition, that they are, therefore, to be classed among uncertain remedies, to be used with great caution.

In India, *gunjah* is used for smoking alone. About 180 grains and a little dried tobacco are rubbed together in the palm of the hand with a few drops of water. This suffices for three persons. A little tobacco is placed in the pipe first, then a layer of the prepared Gunjah, then more tobacco, and the fire above all. Four or five persons usually join in this debauch. The hookah is passed round, and each person takes a single draught. Intoxication ensues almost instantly; from one draught to the unaccustomed, within half an hour; and after four or five inspirations to those more practised in the vice. The effects differ from those occasioned by drinking the *sidhee*. Heaviness, laziness, and agreeable reveries ensue, but the person can be readily roused, and is able to discharge routine occupations, such as pulling the punkah, waiting at table, and divers similar employments.

Young America is beginning to use the bang, so popular among the Hindoos, though in a rather different manner, for young Jonathan must in some sort be an original. It is not a "drink," but a mixture of bruised hemp tops and the powder of the betel, rolled up like a quid of tobacco. It turns the lips and gums a deep red, and if indulged in largely, produces violent intoxication. Lager beer and schnaps will give way for bang, and red lips, instead of red noses, become the "style."

CHAPTER XVIII
SIRI AND PINANG

He took and tasted, a new life
Flowed through his renovated frame;
His limbs, that late were sore and stiff,
Felt all the freshness of repose;
His dizzy brain was calmed,
The heavy aching of his lids
At once was taken off;
For Laila, from the bowers of Paradise,
Had borne the healing fruit.

Thalaba

The widely distributed race of Malays, oc-
cupy not only the Malayan Peninsula, and, though not exclusively,
the islands of the Indian Archipelago, but has penetrated into Mada-
gascar, and spreads itself through the islands of the Pacific from New
Zealand, the Society, the Friendly Isles, and the Marquesas, to the
distant Sandwich and Easter Isles. Whatever may have been the starting point, it is essentially a shore-dwelling race, peopling only islands,
or such portions of the continent as border the ocean, and never penetrating into the interior, or passing the mountains running parallel
to the coast. Their energies are most conspicuous in maritime occupations, and to this predilection their extensive diffusion may be attributed. These people, supposed by some to have an affinity to, or alli-
ance with the Hindoo and Chinese races, whence they have been
called Hindoo-Chinese, present as many points of difference as of resemblance; and while some of the customs of the inhabitants of southern or eastern Asia may be found amongst them, they have also others
peculiarly their own. The indulgence in opium is not unknown to the
Malays, but the national indulgence of the race is the areca or betel
nut, a habit characteristic of a sea-loving people. The use of a pipe,
and especially an opium-pipe, would be a hindrance to the freedom of

194

their motions on board their vessels, and require a state of inactivity
or repose incompatible with a maritime life, in order to be enjoyed.
This may in part account for the prevalence of chewing tobacco in
our navy, and of the "buyo" by the Malays.

The areca palm is one of the most beautiful of the palms of India.
It has a remarkably straight trunk, rising forty or fifty feet, with a diameter of from six to eight inches, of nearly an equal thickness throughout. Six or seven leaves spring from the top, of about six feet in length,
hanging downwards from a long stalk in a graceful curve. This palm is
cultivated all over India, in Cochin-China, Java, and Sumatra, and
other islands of the Archipelago, for the sake of the nuts. The fruit is
of the size and shape of a hen's egg, and consists of an outer, firm,
fibrous rind or husk, about half an inch thick, and an inner kernel,
somewhat resembling a nutmeg in size, but more conical in shape.
Internally the resemblance to a nutmeg, with its alternate white and
brown markings, is even greater. When ripe, the fruit is of a reddish
yellow colour, hanging in clusters among the bright green leaves. If
allowed to hang until fully ripe, it falls off and sows itself in the ground,
but this is not allowed. The trees are in blossom in March and April,
and the fruits may be gathered in July and August, when the sliced
nut can be prepared for them, but they do not fully ripen till September and October.

The nuts vary in size, their quality, however, does not at all depend
upon this property, but upon their internal appearance when cut, inti-
mating the quantity of astringent matter contained in them. If the white
or medullary portion which intersects the red, or the astringent part be
small, has assumed a bluish tinge, and the astringent part is very red,
the nut is considered of good quality, but when the medullary portion is
in large quantity, the nut is considered more mature, and not possessing so much astringency, is not deemed so valuable.

This palm is cultivated in gardens and plantations. The latter are
usually close to the villages, and are extremely ornamental. Like the
Malays themselves, the areca palm prefers the neighbourhood of the
sea, which is most conducive to the perfection of the fruit, as the coca
shrub of the Peruvian mountaineers delights in the slopes of the Andes.
It is stated that a fertile palm will produce, on an average, eight hundred and fifty nuts annually, the average production in the plantation
is about fourteen pounds weight for each palm, or ten thousand pounds

per acre. The price they realize to the grower is about two shillings the hundredweight.

The *addaca*, or betel nut, is a staple product of Travancore. In 1837 the number of trees growing there was stated in the survey to be 10,232,873, which, at the average rate named, would produce 63,000 tons of nuts. Nearly half a million trees are in cultivation in Prince of Wales' Island, which would produce about 3,000 tons annually, exclusive of their supplies from Cochin-China, the amount of which is not known, but, without a doubt, more than another 3,000 tons. Many ships freighted solely with these nuts sail yearly from the ports of Sumatra, Malacca, and Siam.

When there is no immediate demand for the areca nuts they are not shelled, but preserved in the husk, to save them from the ravages of insects, which attack them nevertheless, almost as successfully. Of the nuts produced in Travancore, upwards of 2,000 candies,* prepared nuts, are annually exported to Tinnevelly and other parts of the country, and about 3,000,000 of ripe nuts are shipped to Bombay and other places, exclusive of the quantity consumed in the country, and for the inland trade.

From the report of P. Shungoomry Menowen, we derive the following account of the preparation of the nuts. There are various kinds in use. That used by families of rank is collected while the fruit is tender; the husk, or outer pod, is removed, the kernel, a round fleshy mass, is boiled in water. In the first boiling of the nut, when properly done, the water becomes red, thick, and starch-like, and this is afterwards evaporated into a substance like catechu. The boiled nuts being now removed, sliced, and dried, the catechu-like substance is rubbed thereto, and dried again in the sun, when they become of a shining black colour, and are ready for use. Whole nuts, without being sliced, are also prepared in the same form for use. Ripe nuts, as well as young nuts in the raw state, are used by all classes of people, and ripe nuts preserved in water are also used by the higher classes.

Nuts prepared in Travancore for exportation to Trichonopoly,

---

* 1850—1,734 candies   1853—2,073 candies
  1851—1,983 candies   1854—1,954 candies
  1852—2,953 candies   The candy is 433$\frac{1}{2}$ lbs.

Madura, and Coimbatore, are prepared in thin slices, coloured with red catechu or uncoloured. For Tinnevelly and other parts of the country, the nuts are prepared by merely cutting them into two or three slices and drying them. For Bombay, and other parts of the Northern Country, the nuts are exported in the form of whole nuts dried with the pods.

The nut is chewed by both sexes indiscriminately in Malabar as well as on the Coromandel coast. In Malabar they mix it with betel leaf, chunam, and tobacco; but in Tinnevelly and other parts, tobacco is never added. The three ingredients for the betel, as commonly used, are, the sliced nut, the leaf of the betel pepper, in which the nut is rolled, and chunam, or powdered lime, which is smeared over the leaf.

The areca nut is commonly known by the Malay name of *pinang*, but in the Acheenese language it is called *penu*, and the palm producing it *ba penu*. The ripe nut is called also *penu massa*, and the green *penu mudr*. The leaf of the betel pepper is called either *ranu* or *siri*, and the lime *chunam* or *gapu*. Tobacco, when used, is called *bakun*.

In China, the principal consumption of the nut as a masticatory is in the provinces of Quangton, Quang-se, and Che-keang; and it may be seen exposed for sale on little stalls about the suburbs of Canton with the other additional articles used in its consumption. It is also used in dyeing. In the central provinces of Hoo-kwang and Kang-si the nut is, after being bruised and pounded, mixed with the green food of horses as a preventive against diarrhœa, to which some kinds of food subjects them. The Chinese state that it is used as a domestic medicine in the North of China, some pieces being boiled, and the decoction administered. From them is also prepared a kind of cutch, or catechu, which is exported in great quantities, and is now used largely in this country, together with other kinds, as a tanning and dyeing material.

In Ceylon these instruments are used: the girri for cutting the areca nut, and the wanggedi and moolgah, a kind of mortar and pestle for mincing and intimately mixing the ingredients together.

In Virginia, tobacco was at one period used as a currency at a fixed value per pound. In Peru, the labourer is paid in coca, and in the Philippines, betel rolls have been used in the same manner as currency. To the Malay it is as important as meat and drink, and many would rather forego the latter than their favourite *pinang*. The same

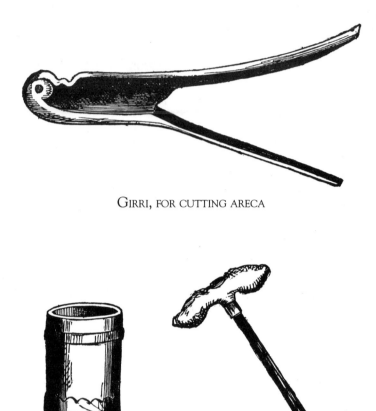

GIRRI, FOR CUTTING ARECA

WANGGEDI OR MORTAR (LEFT) AND MOOLGAH OR PESTLE (RIGHT),
FOR MIXING THE INGREDIENTS

thing might also be said of the inveterate quidder of tobacco; we remember one of this description, who for years used one ounce per day,
and declared often that he had rather be deprived of his dinner than
his quid, although he liked both. Without his leaf, the confirmed
"coquero" is the most miserable of beings, and when deprived of his
customary pipe, the opium-smoker becomes sullen, ill, and utterly incapacitated for his employment. Habits of indulgence of this kind,
when once commenced, are not so easily thrown off. It has been said
that a "coquero" was never reclaimed from the use of his coca.

No estimate can be given of the absolute quantity of areca nuts which are used as a masticatory. Johnston calculates that they are chewed by not less than fifty millions of people, which, at the rate of ten pounds per year, or less than half an ounce per day, would amount to two hundred and twenty thousand tons, or five hundred millions of pounds, a quantity greater than that of any other narcotic except tobacco.

Areca nuts have been strung and made into walking sticks,* and, in this country, turned and formed into ornamental bracelets, as well as burnt into charcoal for tooth powder. We have engirdled the earth with pig-tail, let us apply the same kind of calculation to the estimated annual consumption of areca nuts, and strung together in the form of a bracelet, we have a string 505,050 miles in length, enough to go round the world 21 times; or, supposing these nuts to be arranged side by side, they would cover a road fourteen feet wide for the distance of not less than 3,000 miles. If arranged in like manner in the form of a square, they would occupy at least 5,000 acres of land.

The areca palm has given its name to the island of Penang, not from its growing there in larger numbers, or more luxuriant than elsewhere, but because it was the tree chiefly cultivated by the Malays who first occupied the island. It now better deserves the title, being the emporium for the betel nut raised on the east coast of Sumatra.

In Sumatra many of the common drinking and baking utensils in the boats, and vessels for holding water, not dissimilar to those made by the Australian natives from the bark of the gum trees, are made from the spathe of this palm, it is also nailed upon the bottoms of the boats, and often small bunches of the abortive fruit may be seen placed as an ornament at the stem and bows of the native vessels. The male flowers are deliciously fragrant, and are in request in the island of Borneo on all festive occasions; they are considered a necessary ingredient in the medicines and charms employed for healing the sick. In Malabar an inebriating lozenge is prepared from the sap of this palm.

Manuel Blanco thinks that the areca might be used for making red ink, and it is not improbable that it is thus employed in India. With other combinations it makes black ink of moderate quality. The

---

*There is a stick of this kind in the Museum of Economic Botany at Kew Gardens.

lower part of the petiole is used for wrapping instead of paper, for which purpose it is sold in the Philippines. The heart of the leaves is eaten as a salad, and has not a bad flavour. The convicts confined in the Andaman Islands masticate the nuts of another species of areca. The Nagas and Abors of Eastern Bengal, use those of a third species, and the natives of the mountainous districts of Malabar those of a fourth. There are about twenty species of the areca genus, of which several are thus used.

When betel nuts are scarce in the Philippines, the natives substitute the bark of the Guayabo and the Antipolo.

It is confidently affirmed to us, that in Ceylon the natives sometimes masticate the roots of the cocoa-nut palm, instead of, and as a substitute for, the areca nut, and that it answers the purpose very well.

The root of a plant known botanically as *Derris pinnata*, is also occasionally used amongst certain Asiatics, in the same manner, in cases of deficiences in the supply of genuine betel.

The consumption of the areca nut being confined to an area of no very wide extent, and that principally in the neighbourhood of the producing countries, or *in* those countries themselves, the necessity for providing a substitute does not often arise; hence, those of which we have any knowledge, as having been at all generally used for that purpose, are confined to two or three substances. Some years, however, are not so productive as others, and instances have occurred in which the average price of areca nuts for mastication has been doubled. If the Yankees persist in their betel and hemp chewing propensities, which have lately been developed amongst them, probably the Chinese and Malay will have to pay a higher price for their nuts, or provide something which shall thenceforth fulfil its duties, and we may hear of other substitutes.

Ardent as the admirers of the areca may be in their admiration of the "buyo," we have never seen more than one translation of a Malayan poem in which the masticatory was extolled, and this, unfortunately, we are unable to present to our readers. The gods have either not made the votaries of betel so poetical as the servants of the pipe, or the peans in praise thereof are locked up from us in the cabalistic characters of their national language. The unmistakable marks left by the habit on the lips, teeth, and gums, are certainly extolled by them as marks of beauty. In the poem already referred to, the lover addresses his mistress in praise of the redness of her teeth and lips, and the fragrant odour of her breath, produced by the sweet "buyo" secreted in the hollow of her cheek. White teeth are therefore held in abomination, and as this is also the opinion of certain African tribes, who stain theirs with the juice of flowers, ours *must* be a barbarous nation to respect such albino masticators.

N.B.—The average annual export of areca nuts from Ceylon is 50,000 cwts., and the price a fraction below 20s. per cwt.

CHAPTER XIX

UNDER THE PALMS

A wind blew warm from the east, and it lifted its arms hope-
lessly; and when the wind, love-laden with most subtle sweet-
ness, lingered, loth to fly, the palm stood motionless upon its
little green mound, and the flowers were so fresh and fair, and
the leaves of the trees so deeply hued, and the native fruit so
golden and glad upon the boughs, that the still warm garden air
seemed only the silent, voluptuous sadness of the tree; and had I
been a poet my heart would have melted in song for the proud,
pining palm.

G. W. Curtis

Two species of a kind of pepper vine are
extensively cultivated, with the areca palm, in all the countries of the
East where chewing the betel is indulged in. These belong to the same
family of plants as those producing the common black pepper and the
long pepper of commerce. They are known to botanists as *Chavic betle*
and *Chavica siraboa*. They are similar in their habits, being trailing
plants, with some resemblance to the ivy, but more tender and fragile.
The betel palms may be often seen with the pepper, climbing and
twining around their tall, straight, slender trunks, or they are trained
about poles of bamboo in the manner of hops in the hop gardens of
Kent. Almost every one with a piece of land cultivates the pepper for
his own consumption. In the markets incredible quantities of the leaves
are offered for sale, in piles carried about in baskets. In Northern India, sheds are constructed for the growth of the pepper. These are
from twenty to fifty yards in length, and eight or twelve broad, of
bamboo, to shelter the plants from the sun. Great attention is paid to
the cultivation, and the plants are carefully attended to, and cleaned
every morning.

Betel leaf cannot be preserved in a sound state beyond eight days
without preparation, but by being prepared over a fire, and rolled into

balls, in which state it is called *chenai*, it will keep a year, only the quality is much deteriorated. In Penang the old men carry about with them a sort of metal tube, having a ramrod-looking pestle, with which they busy themselves in pounding the mixture for chewing. The young daily make nut-crackers of their jaws, and although the mixture, perhaps, rather tends to preserve the teeth, still the exercise on the nut must be a little too violent for them, and the Malays say it injures the sight. The Chinese are not much addicted to the use of the betel.

The consumption of betel by the inhabitants of Penang and Province Wellesley may be stated at 6,211,440 bundles of 100 leaves each, equal in value to 31,057 Spanish dollars, which would be the produce of 98 orlongs of land, or about 130 acres, planted regularly. But allowing for the various distances given by different cultivators between the plants 110 orlongs may be assumed, or about 147 acres.

The Chinese colonists of Singapore used the leaves of the common pepper, instead of those of the betel pepper in compounding this masticatory.

The Ava pepper, or *Macropiper methysticum*, is even more celebrated for its narcotic properties than the two just referred to. This plant has a thick aromatic wood stalk, and a large root, and cordate or heart-shaped leaves. It is a native of the Society, Friendly, and Sandwich Islands, where it is largely consumed. Macerated in water, the stems and root form an intoxicating beverage, and the leaves are used with the areca nut and lime, in the same manner as the leaves of the other peppers.*

Mariner gives an account, in his *History of the Tonga Islands*, of the use of this plant. The root is split up with an axe into small pieces, and after being scraped clean with mussel shells, is handed out to those in attendance to be chewed. There is then a buzz in the assembly, contrasting curiously with the silence which reigned before, several crying out, "Give me some cava! give me cava," each of those who intend to chew it crying out for some to be handed to him. No one offers to chew the cava but young persons who have good teeth, clean

---

* The stem and roots of long pepper, cut in pieces and dried under the name of *Pipula moola*, are exposed for sale in all the bazaars of India, but these are not used with the areca nut, nor are the leaves applied to that purpose.

mouths, and no colds. Women frequently assist. It is astonishing how remarkably dry they preserve the root during the process of mastication. In about two minutes, each person having chewed his quantity, takes it out of his mouth with his hand, and puts it on a piece of plantain or banana leaf, or he raises the leaf to his mouth, and puts it off from his tongue, in the form of a ball of tolerable consistence. The different portions of cava being now chewed, which is known by the silence that ensues, a large wooden bowl is placed on the ground before the man who is to make the infusion. Each person passes up his portion of the chewed root, which is placed in the bowl, wherein they are laid in such a manner that each portion is distinct and separate from the rest, till the whole inside of the bowl becomes studded, from the bottom up to the rim, on every side. The man, before whom the bowl is placed, now tilts it up a little towards the chief, that he may see the quantity of its contents, saying, "This is the cava chewed." If the chief thinks there is enough, he says, "Cover it over, and let there come a man here." The bowl is covered over with a plantain or banana leaf, if there is not enough, and a man fetches more root to be chewed. If there is enough, the chief says "mix." The two men, who sit on each side of him, who is to prepare the cava, now come forward a little, and making a half turn, sit opposite to each other, the bowl being between them, one of these fans off the flies with a large leaf, while the other sits ready to pour in the water from cocoa-nut shells, one at a time.

Before this is done, however, the man who is about to mix, having first rinsed his hands with a little of the water, kneads together the chewed root, gathering it up from all sides of the bowl, and compressing it together. Upon this an attendant says, "Pour in the water," and the man on one side of the bowl continues pouring, fresh shells being handed to him, until the attendant thinks there is sufficient, and says, "Stop the water." The mixture is stirred together at the command of the attendant, who then says, "Put in the fow," which is the bark of a tree stripped into small fibres, and has the appearance of willow shavings. A large quantity of this substance, enough to cover the whole surface of the infusion, is now put in by one of those seated beside the bowl, and it floats upon the surface. The man who manages the bowl now begins his difficult operation. In the first place, he extends his left hand to the further side of the bowl, with the fingers pointing downwards and the palm towards himself; he sinks that hand carefully down the side of the bowl, carrying with it the edge of the fow; at the same time his right hand is performing a similar operation at the side next to him, the fingers pointing downwards and the palm presenting outwards. He does this slowly from side to side, gradually descending deeper and deeper, till his fingers meet each other at the bottom, so that nearly the whole of the fibres of the root are by these means enclosed in the fow, forming, as it were, a roll of about two feet in length, lying along the bottom from side to side, the edges of the fow meeting each other underneath. He now carefully rolls it over, so that the edges overlapping each other, or rather intermingling, come uppermost. He next doubles in the two ends and rolls it carefully over again, endeavouring to reduce it to a narrower and firmer compass. He now brings it cautiously out of the fluid, taking firm hold by the two ends, and raising it breast high, with his arms extended; by a series of movements the mass is more and more twisted and compacted together, while the infusion drains from it in a regular decreasing quantity, till, at length, it denies a single drop. He now gives it to the person on his left side and receives fresh fow from the one on the right. The operation is again renewed with a view to collect what might before have escaped him, and even a third time till no dregs are left which this process can remove.

During the above operation, various people are employed in making cava cups from the unexpanded leaves of the banana, folded and tied in a peculiar manner. The infusion being strained, the performance generally occupying a quarter of an hour or twenty minutes, the man at the bowl calls out, "The cava is clear." The infusion is now filled into the cups by means of a bundle of fow which is dipped into the bowl, and when replete with the liquid, held over the cup, and being compressed, the liquid runs out till the cup is filled. With certain other ceremonies the cups are passed round amongst the company.

From this account it will be seen that the beverage is drank immediately after it is prepared, without being in any manner fermented, its intoxicating and narcotic properties must, therefore, be due to the root. This liquor is indulged in to a large extent in the islands of Oceanica, where the natives are generally passionately fond of it.

Another substance entering into the composition of the "buyo" is the extract of the leaves of the gambir *(Uncaria gambir)*. There are different qualities of extract: the first and best is white, brittle, and has an earthy appearance when rubbed between the fingers, which earthy appearance gave it the name of Terra Japonica, being supposed, at first, also, to come from Japan, and is formed into very small round cakes. This is the most expensive kind, and most refined, but it is not unfrequently adulterated with sago; this kind is brought in the greatest quantity from the island of Sumatra. The second quality is of a brownish yellow colour, is formed into oblong cakes, and when broken has a light brown earthy appearance; it is also made into solid cubic form; it is sold in the bazaars in small packets, each containing five or six. The third quality contains more impurities than the preceding, is formed in small circular cakes, and sold, in packages of five or six, in the bazaars.

The method employed in making the extract is thus described in the *Singapore Chronicle*:

> The leaves are collected three or four times a year; they are thrown into a large cauldron, the bottom of which is formed of iron, the upper part of bark and boiled for five or six hours, until a strong decoction is inspissated, it is then allowed to cool, when the extract subsides. The water is drawn off, a soft, soapy substance remains, which is cut into large masses; these are further divided by a knife into small cubes, about an inch square, or into still smaller pieces, which are laid in frames to dry. This catechu has more of a granular uniform appearance than that of Bengal, it is, perhaps, also less pure. The younger leaves of the shrub are said to produce the whitest and best gambir, the older a brown and inferior sort. The men employed in the gambir plantations generally indulge freely in the use of opium.

Another extract made in India from the wood of *acacia catechu*,* and which bears the name of Cutch or "Kutt," is used in combination with the betel nut. The trees are cut down, and the heart-wood chopped and boiled in water, strained off, and evaporated. This is poured into clay moulds and dried in the sun. Dr. Hooker gives a sketch from the life of one of the native "Kutt" makers of India:

---

* From *cate* a tree, and *chu* juice.

At half-past eight A.M. it suddenly fell calm, and we proceeded to Chakuchee, the native cars breaking down in their passage over the projecting beds of flinty rocks, or as they hurried down the inclined planes which cut through the precipitous banks of the streams. Near Chakuchee we passed an alligator, just killed by two men—a foul beast about nine feet long, and of the Mager kind. More interesting than its natural history was the painful circumstance of its having just swallowed a child that was playing in the water, while its mother was washing her domestic utensils in the river. The brute was hardly dead, much distended by its prey, and the mother standing beside it. A very touching group was this! the parent with hands clasped in agony, unable to withdraw her eyes from the cursed reptile, which still clung to life with that tenacity for which its tribe is so noted, and beside her the two men leaning on their bloody bamboo staves with which they had all but despatched the animal.

The poor woman who had lost her child earns a scanty maintenance by making catechu. She inhabits a little cottage, and has no property but her two oxen to bring wood from the hills, and a very few household chattels, and how few these are is known only to persons who have seen the meagre furniture of the Dangha hovels. Her husband cuts the trees in the forest and drags them to the hut, but he is now sick, and her only son, her future stay, was he whose end is just related. Her daily food is rice, with beans from the beautiful flowered dolichos, trailing round the cottage, and she is in debt to the contractor, who has advanced her two rupees, to be worked off in three months, by the preparation of 240 lbs. of catechu. The present was her second husband, an old man; by him she never had any children, and in this respect alone did the poor creature think herself very unfortunate, for her poverty she did not feel. Rent to the Rajah, tax to the police, and rates to the Brahminee priest, are all paid from an acre of land, yielding so wretched a crop of barley, that it more resembled a fallow field than a harvest field. All day long she is boiling down the decoction into wooden troughs, where it is inspissated.

From the areca nut another kind of catechu is prepared, which is generally preferred as a masticatory. Heyne thus describes the process of its manufacture, "Areca nuts are taken as they come from the tree,

and boiled for some hours in an iron vessel. They are then taken out
and the remaining water is inspissated by continued boiling. This process furnishes *kassu*, or most stringent *terra japonica*, which is black,
and mixed with paddy husks and other impurities. After the nuts are
dried, they are put in a fresh quantity of water, boiled again, and the
water, being inspissated like the former, yields the best or dearest kind
of catechu, called *coury*. It is yellowish brown, has an earthy fracture,
and is free from the admixture of foreign bodies." It is probable that
the flat round cakes, covered with paddy husks, met with in commerce is the *kassu* of Heyne.

The husk which surrounds the nut, and which is of a fibrous nature, resembling the coir of the cocoa-nut is thrown away by tons, and
allowed to rot. This substance has lately been experimented upon for
the manufacture of paper, for which purpose it appears to be available,
and, as there is no want of the raw material, perhaps at some future
time it will become utilized as extensively as the "coir" of Ceylon.

The Bombay catechu is obtained from *Acacia catechu*, and the
Bengal catechu from *Uncaria Gambir*. The Bombay produce is of a
dark brownish red colour, and is stated to be the richer of the two in
tannin. The Bombay variety is commonly called "cutch," while the
Bengal produce is of a lighter brown colour, and is termed "terra."
Catechu of good quality is also obtained from Peru.

The catechu exported from Madras to England, Bombay, France,
and Ceylon was—

1853–4	484 cwt.	valued at	£199	4s.
1854–5	1,364		698	8
1855–6	2,908		2,297	2
part of 1856–7	658		270	8
Or in 3¹/₂ years	5,414 cwt.	valued at	£4,265	2

But this is only a small proportion of the catechu consumed in
England alone, since in 1849 we imported 169,140 cwts. of that substance for tanning purposes, and the quantity has since increased.

The totals of cutch and gambier imported in

1856 was 8,536 tons.
1857 was 11,047 tons.
1858 was 11,205 tons.
1859 was 13,762 tons.

Of this quantity we exported in—

1856	1,031 tons.
1857	1,427 tons.
1858	974 tons.
1859	1,809 tons.

These articles, therefore, make no insignificant item in our East Indian trade, which, valued at the intermediate rate of 15s. and 30s. per cwt., would amount to the sum of £153,375 in 1858.

CHAPTER XX
CHEWING THE COON

It ascends me into the brain, dries me there all the foolish,
and dull, and crudy vapours which environ it; makes it
apprehensive, quick, forgetive, full of nimble, fiery, and
delectable shapes, which, delivered over to the voice (the
tongue), which is the birth, becomes excellent wit.

Sir John Falstaff

"In Burmah," says Howard Malcolm, "almost every one, male and female, chews the singular mixture called *coon*, and the lacquered or gilded box containing the ingredients is borne about on all occasions. The quid consists of a slice of areca nut, a small piece of cutch, and some tobacco rolled up in a leaf of betel pepper, on which has been smeared a little tempered quicklime. It creates profuse saliva, and so fills up the mouth that they seem to be chewing food. It colours the mouth deep red, and the teeth, if not previously blackened, assume the same colour. From the combination of the three ingredients this colour seems to proceed, since the leaf and nut, without the lime, fail to produce it. This hue, communicated to the mouth and lips, is esteemed ornamental, and an agreeable odour is imparted to the breath. The juice is usually, though not always, swallowed. A curious circumstance connected with the expectoration of the red juice is related at Manilla, where it is narrated with strong protestations and firm belief in its veracity.

Some years ago a ship from Spain arrived in the port of Manilla. Among the passengers was a young doctor from Madrid, who had gone to the Philippines with the design of settling in the colony and pushing his fortune by means of his profession. On the morning after he had landed, our doctor sallied forth for a walk on the pasco. He had not proceeded far when his attention was attracted to a young girl, a native, who was walking a few paces ahead of him. He observed that every now and then the girl stooped her head towards the pavement which was straightway spotted with blood. Alarmed on the girl's account, our doctor walked rapidly after her, observing that she still continued to expectorate blood at intervals as she went. Before he could come up with her the girl had reached her home, a humble cottage in the suburbs, into which she entered. The doctor followed close upon her heels, and summoning her father and mother, directed them to send immediately for the priest as their daughter had not many hours to live. The distracted parents, having learned the profession of their visitor, immediately acceded to his request. The child was put to bed in extreme affright, having been told what was about to befall her. The nearest padre was brought, and everything was arranged to smooth the journey of her soul through the passes of purgatory. The doctor plied his skill to the utmost, but in vain. In less than twenty-four hours the girl was dead.

As up to that time the young Indian had always enjoyed excellent health, the doctor's prognostication was regarded as an evidence of great and mysterious skill. The fame of it soon spread through Manilla, and in a few hours the newly-arrived physician was beleagured with patients, and in a fair way of accumulating a fortune. In the midst of all this, some one had the curiosity to ask the doctor how he could possibly have predicted the death of the girl, seeing that she had been in perfect health a few hours before. "Predict it," replied the doctor, "why, sir, I saw her spit blood enough to have killed her half a dozen times."

"Blood! how did you know it was blood?"

"How! from the colour, how else?"

"But every one spits red in Manilla."

The doctor, who had already observed this fact, and was labouring under some uneasiness in regard to it, refused to make any further confession at the time, but he had said enough to elucidate the mystery. The thing soon spread throughout the city, and it became clear to every one that what the new *medico* had taken for blood, was nothing else than the juice of the buyo, and that the poor girl had died from the fear of death caused by his prediction. His patients now fled from him as speedily as they had congregated; and to avoid the ridicule that awaited him, as well as the indignation of the friends of the deceased girl our doctor was fain to escape from Manilla, and return to Spain in the same ship that had brought him out.

· The ladies who work in the government cigar factory at Manilla, all, more or less, chew the betel nut, and any one daring enough to disregard the warning not to touch anything, when passing as a visitor through the rooms, must stand the assault from the mouths of a hundred or two of these dames, in the shape of a deluge of the decoction of this nut. The captain of an American vessel at Manilla, although warned of the consequences, with American impudence, infringed the rule, and paid the penalty. He was compelled to beat a retreat, and being dressed in the white garb of the East, resembled a spotted leopard, in the room of a free and enlightened citizen of the great Republic.

The mastication of the betel is considered very wholesome by those who are in the habit of using it, and it may be so, but the black appearance it gives to the teeth, although it is said to be an excellent preserver of them, together with the brick red lips and mouth, cause anything but an agreeable appearance. Its use certainly does not impart additional beauty to the native females, who habituate themselves to an extent equal to that of the opposite sex.

The custom, Marsden states, is universal among the Sumatrans, who carry the ingredients constantly about them, and serve them to their guests on all occasions; the prince in a gold stand, and the poor man in a brass box or mat bag. The betel-stands of the better ranks of people are usually of silver, embossed with rude figures. The Sultan of Moco-Moco was presented with one by the India Company with their arms upon it, and he possesses another besides, of gold filagree. The form of the stand is the frustum of an hexagonal pyramid, reversed, about six or eight inches in diameter. It contains many smaller vessels, fitted to the angels, for holding the nut, leaf, and chunam, with places for the instruments employed in cutting the first, and spatulas for spreading the last.

Captain Wilkes also describes that of the Sultan of Sooloo. "On the left hand of the Sultan sat his two sons, on the right his councillors, while immediately behind him sate the carrier of his betel-nut casket. The casket was made of filagree silver, about the size of a small tea-caddy, of oblong shape, and rounded at the top. It had three divisions, one for the nut, another for the leaf, and a third for the lime. Next to this official was the pipe-bearer, who did not appear to be held in equal estimation."

A circumstance is also narrated in connection with the son of this same Sultan, which exhibits the use of betel in another phase. This son, shortly after taking a few whiffs from the opium-pipe was overcome, and became stupid and listless. When partially recovered, he called for his betel nut to revive him by its exciting effects, and counteract the influence of the opium. The pinang or buyo was carefully chewed by his attendant to a proper consistency, moulded into a ball, and then slipped into his mouth. Hence we may learn two things. First, that chewing the betel counteracts the ill effects of an over-dose of opium. Secondly, that it is extremely convenient to have an attendant with a good set of teeth, since he could not only masticate betel nut for you, and relieve you from a large amount of labour; but in the event of your joint not being so tender as it should be, the amount of milling to be expended at dinner could be divided between you, the attendant masticating the tough dishes, and yourself the tender, and thus, by division of labour, a good dinner could be procured with little expenditure of your own muscular strength.

In Sumatra, when the first salutation is over, which consists in bending the body, and the inferior putting his joined hands between those of the superior, and lifting them to his forehead, the betel is presented as a token of hospitality and an act of politeness. To omit it on the one hand, or to reject it on the other would be an affront, as it would be, likewise, in a man of subordinate rank to address a great man without the precaution of chewing it before he spoke.

The Tagali maidens, says Meyen, regard it as a proof of the uprightness of the intentions of a lover, and of the strength of his affection, if he takes the buyo from his mouth. In Luçon, a little box or dish is kept in every house, in which are kept the betel rolls prepared for the day's consumption, and there a buyo, or betel roll, is offered to every one who enters, just as a pinch of snuff or a pipe might be with us. Making the buyo is a part of the occupation of the females, who may be seen in the forenoon stretched on the ground rolling them. Enough for the day's consumption is generally carried, in a siri box of metal or japanned ware, by those whose occupation call them from home; every one who can afford the expense, puts a fresh roll in his mouth every hour, which he continues to chew and suck for about half an hour or more.

Betel holds an important place in the marriage ceremonies of the Tagals. When once a young man has informed his father and mother that he has a predilection for a young Indian girl, his parents pay a visit to the young girl's parents upon some fine evening, and after some very ordinary chat, the mamma of the young man offers a piastre to the mamma of the young lady. Should the future mother-in-law accept, the young lover is admitted, and then his future mother-in-law is sure to go and spend the very same piastre in betel and cocoa wine. During the greater portion of the night, the whole company assembled upon the occasion, chews betel, drinks cocoa wine, and discusses upon all other subjects but marriage. The young men never make their appearance till the piastre has been accepted, because in that case they look upon it as being the *avant-courier*, that is, the first and most essential step towards their marriage.

During the fast of Ramadan, the Mahometans abstain from the use of the betel while the sun continues above the horizon, but, except at this time, it is the constant luxury of both sexes from an early period of childhood till old age, when, becoming toothless, they are unable to masticate the nut, and are reduced to the alternative of having all the ingredients previously reduced to the form of a paste for them, so that, without effort, they may dissolve in the mouth.

When Lady Raffles had reached Merambung in Sumatra, being much fatigued with walking, and the rest of the party having dispersed in various directions, she lay down under the shade of a tree, when a Malay girl approached, with great grace of manner, and on being asked if she wanted anything, replied, "No! but as you were quite alone, I thought you might like to have a little talk, so I came to offer you some *siri* (betel), and sit beside you."

The darker the teeth the more beautiful is a Siamese belle considered; and in order that their gums should be of a brilliant red, to form a pleasant contrast to the black lips and teeth, they resort to the pastime of chewing betel from morning till night. The constituents of the betel being rolled up into something very much like a sailor's quid, it is then thrust into the lady's cheek, and is munched, and crunched, and chewed so long as the slightest flavour is to be extracted, and, as they never swallow the juice, the results are very detrimental to the cleanliness of the floors of the houses, and of themselves generally. They commonly make use of two such quids during the day, and this mixture has the effect of dyeing their gums and the whole of the palate and tongue of a blood-red colour. Old crones, and very ancient *chronoses* (for both men and women use the betel), who have no longer any teeth to masticate the mixture with, are attended by servants, who have a species of small pestle and mortar always about them, wherein they reduce the betel into a proper form for the delicate gums of their aged patrons.*

The betel pepper is cultivated at Zanzibar, where the use of betel prevails, as it does at the Comoro Islands and at Bombay. But the custom is not in vogue in Arabia. The betel palm is also grown for the sake of its fruits in the island of Zanzibar.

The habit of masticating betel nut in combination with hemp has of late come into vogue in the United States of America, and doubt-less Brother Jonathan will soon eclipse Malaya in his predilection for the "buyo."

---

* Neale's residence in Siam.

CHAPTER XXI

OUR LADY OF YUNGAS

And all my days are trances;
And all my nightly dreams
Are where thy dark eye glances,
And where thy footstep gleams:
In what etherial dances,
By what eternal streams.

E. A. Poe

To the Peruvian the province of Yungas de la Paz in the North-East of Bolivia is an El Dorado, because *there* grows in the greatest profusion and luxuriance his favorite coca. We may look with delight towards the island of Ceylon, and, in imagination, snuff the fragrant breezes that have passed over the cinnamon groves and coffee plantations; or direct the gaze of our children cross the map of the world to South-Eastern China, and inform them that from thence to the United States, and add that from this place their worthy sires receive the greater part of their tobacco. But the affections of the Peruvian are not so divided; they are located upon one spot, and *that* the province of the "warm valleys," or the Yungas de la Paz; there dwells his patron saint, and from thence *he* receives the "keys of Paradise."

At the time of the conquest the coca was only used by the Incas, and those of the royal, or rather solar, blood. It was cultivated for the monarch and for the solemnities of their religion; none might raise it to his mouth, unless he had rendered himself worthy by his services to partake of this honour with his sovereign. The plant was looked upon as an image of divinity, and no one entered the enclosures where it was cultivated without bending the knee in adoration. The divine sacrifices made at that period were thought not to be acceptable to Heaven,

216

unless the victims were crowned with branches of this tree. The oracles made no reply, and auguries were terrible if the priest did not chew *coca* at the time of consulting them. It was an unheard of sacrilege to invoke the shades of the departed great without wearing the plant in token of respect, and the Coyas and Mamas who were supposed to preside over gold and silver, rendered the mines impenetrable unless propitiated by it. In the course of time its use extended, and gradually became the companion of the whole Indian population. To this plant the native recurred for relief in his greatest distress; no matter whether want or disease oppressed him, or whether he sought the favours of Fortune or Love, he found consolation in the "divine plant."

The word by which this plant is known has been referred, for its etymology, to the Aymara language, in which *Khoka* signifies *tree* or *plant*. It is known that the shrub producing the Matè or Paraguay tea, the favourite beverage of many South American nations, is called *la Yerba*, i.e. *the plant*. As also in Mexico tobacco was called *yetl*, and by the Peruvians *Sagri*, meaning in those languages *the herb*, so we, occasionally, are apt to designate the latter article *the weed*. Showing, that to those persons or nations who have appropriated such names, trivial in themselves, to the different articles of consumption, these plants were in themselves pre-eminent in the vegetable creation, as, in another instance, we have shown our appreciation of one book above all others, century after century, by the simple designation of *The Book*.

In Europe, the historians of the conquest gave the first information of the sacred plant of the Peruvians; this was, however, merely superficial. In 1569, Monardes, and in 1605 Clusius, wrote concerning it, but the leaves of the plant itself were not seen until brought over by one of he companions of La Condamine, Joseph de Jussieu, who nearly lost his life in 1749, while crossing the Cordilleras in search of this plant. He was compelled to cross the mountains, covered as they were with snow, on foot, descending by means of paths cut out like ladders, and overhanging frightful precipices. The intensity of the sun's rays, reflected by the snow, caused him the most distressing pains in the eyes, and almost blinded him, but the success of his expedition consoled him for the misfortunes that he had endured.

This shrub rises to the height of from four to eight feet, the stem covered with whitish tubercles, which appear to be formed of two curved lines set face to face. The leaves are oblong, and acute at each end, from an inch and a half to two inches in length. The leaves are the only parts used, for which purpose they are collected and dried. The shrub is found wild in Peru, according to Pöppig, in the environs of Cuchero, and on the stony summit of the Cerro de San Christobal. It is cultivated extensively in the mild, but very moist climate of the Andes of Peru, at from 2,000 to 5,000 feet above the sea level; in colder situations it is apt to be killed, and in warmer to lose the flavour of the leaf.

The coca plant is propagated from seed sown in nursery beds and carefully watered. When about sixteen or eighteen inches high they are transplanted into plantations called *cocals*, in terraces upon the sides of the mountains. At the end of a year and a half the plant affords its first crop, and from this period to the age of forty years or more it continues to yield a supply. Instances have been noticed of coca plantations that have existed for near a century; but the greatest abundance of leaves is obtained from plants between the third and sixth years. There are four gatherings in the season; the first takes place at the period of flowering, and consists of the lower leaves only. These are larger and less finely flavoured than those afterwards collected, and are mostly consumed at once. The next and most abundant harvest takes place in March; the third and most scanty, in June or July, and the last in November. The leaves are collected similarly to those of tea. Women and children are employed for this purpose. The gatherer squats down, and holding branch with one hand, plucks from it the leaves, one by one, with the other. These are deposited in a cloth, from which they are afterwards collected into sacks to be conveyed from the plantation. The sacks of leaves are carried to the *haciendas*, where they are spread upon a floor of black slate to dry in the sun. They are then packed up in bales made of banana leaf, closely pressed together, each bale containing on an average twenty-four pounds. The price realised to the cultivator is one shilling per pound.

Dr. Weddell endeavoured to obtain reliable information as to the quantity of coca cultivated and collected in the province of Yongas, and states, as a result, that the annual produce is about 400,000 bales, or 9,600,000 Spanish pounds. There is also a large cultivation, not only in other parts of Bolivia, and in Peru, but also in parts of Brazil, so that this cannot represent more than half the amount of the annual consumption of coca. It is true that Pöppig estimated fifteen millions of pounds as the quantity consumed, but this would be too small. On the other hand, Johnston estimates the consumption at thirty millions of pounds; this is, probably, erring rather on the contrary side. Of this quantity he estimates the value at one million and a half sterling, and concludes that the chewing of coca is indulged in by about ten millions of the human race. This again is rather a "long bow"; the use of coca seems to be confined to Peru, Bolivia, and Brazil—at any rate, it is confined to South America, and there is no mention of its indulgence in Chili to the South, or in the Columbian Republics to the North. It would, moreover, confer upon us somewhat of a personal favour, were some one to convince us that the male population of South America amounts to the number which the professor has estimated as that of the indulgers in coca. Our own impression is, that the entire population has only been estimated at seventeen and a quarter millions: this is, at least, the mean of four very respectable authorities. Suppose half of these to be children, and half of the residue females, and we have only an adult male population of less than four and a half millions in the southern half of the New World. Ye shades of Cocker and De Morgan! tell us how from these we can subtract ten millions who indulge in coca, and yet show a remainder, be it ever so small, of abstainers. But it has never been affirmed that coca was indulged in, except in Peru, Bolivia, and Brazil. The population of these three countries amount, according to the higher authorities, only to ten millions, so that every man, woman, and child, must be a coquero to reach the estimated number. Viewing this subject in another of its phases—Johnston states that the average consumption of the coquero is from one ounce to one ounce and a half per day, or, according to ordinary computation, twenty-two to thirty-three pounds per year, whereas the estimated production, which we have presumed to be too large, is, in fact, too small for the number estimated as indulging therein, as it only allows each coca masticator three pounds per annum. In all deference to so high an authority, we will venture to suggest that were the number indulging in coca limited to two millions, and the supply to twenty millions of pounds, or ten pounds annually to each person, some of these difficulties would be removed; but, out of regard for the patience of our readers, we will forbear detailing any further calculations, or the bases on which they rest.

At first the Spaniards strenuously opposed the use of the coca—it was anathematized by them everywhere, as tobacco was by its zealous

opponents in the Old World, but this opposition only seemed to produce an extension of the habit. Then the Spaniards, appreciating the
advantages which might accrue to them in a monopoly of the plant,
took the culture into their own hands, and by force, enrolled the Indians of the Cordilleras in their service, much to the discomfort of the
latter, who suffered extremely from the change of climate. Complaints
to the government being so numerous, the Viceroy, Don Francisco de
Toledo, espoused the cause of the Indians, published seventy-one de-
crees in their favour, and the speculation was abandoned. It is said, that
in 1583 the government of Potosi derived a sum not less than £100,000
from the consumption of 90,000 to 100,000 baskets of this leaf. The
cultivation of coca is therefore an important feature in Peruvian hus-
bandry, and so lucrative, that a coca plantation, whose original cost
and current expenses amounted to £500 during the first twenty months,
will, at the end of ten months more, bring a clear income of £340.

The coca possesses a slightly aromatic and agreeable odour, and
when chewed, dispenses a grateful fragrance, its taste is moderately
bitter and astringent, and somewhat resembles green tea; it tinges the
saliva of a greenish hue. Its effects on the system are stomachic and
tonic, and it is said to be beneficial in preventing intermittents, which
have always prevailed in this country.

The mode of employing coca is to mix with it in the mouth a
small quantity of lime prepared from shells, much after the manner
that the betel is used in the East. With this, a handful of parched
corn, and a ball of arrow-root, an Indian will travel on foot a hundred
leagues, trotting on ahead of a horse. On the frequented roads, we are
informed, that the Indian guides have certain spots where they throw
out their quids, which have accumulated into little heaps, that now
serve as marks of distance; so that, instead of saying, one place is so
many leagues from another, it is common to call it so many quids. Dr.
Weddell states that the Bolivians are in the habit of using instead of
lime with their leaf, a substance called *llipta*, which consists of the
ashes of the Quinoa plant; in other parts the ashes of other plants are
used, as on the Amazon, those of the leaves of the trumpet-tree. These
alkaline ashes are made into little cakes, and sold in the markets.

The Peruvian ordinarily keeps his coca in a little bag called *chuspa*,
which he carries suspended at his side, and which he places in front

whenever he intends to renew his *chique*, which he does at regular intervals, even when travelling. The Indian who prepares himself to chew, in the first place sets himself as perfectly at ease as circumstances permit. If he has a burden, he lays it down; he seats himself, then putting his *chuspa* on his knees, he draws from it, one by one, the leaves which are to constitute his fresh "quid." The attention which he gives to this operation is worthy of remark. The complaisance with which the Indian buries his hand in the leaves of a well-filled *chuspa*, the regret he seems to experience when the bag is nearly empty, deserve observation, for these little points prove that to the Indian the use of coca is a real source of enjoyment, and not the simple consequence of want.

We remember an elderly lady* who was in the habit of taking snuff with the same amount of ceremony. First, she comfortably seated herself, arranged her dress, and smoothed her apron. The most important occupations always being for the time put aside, and apparently forgotten. The next operation consisted in drawing from some capacious receptacle, the entrance to which was enveloped in the folds of her outer garment, a large brown handkerchief, studded with small yellow spots, just visible, we remember it for years, and never any other; this was laid upon the lap prepared to receive it. Another step consisted in drawing out from the same mysterious receptacle, a black japanned box, circular in shape, and of the diameter of a shaving-box, but scarce an inch in thickness; this was carefully wiped with the handkerchief already named, and then grasped in the left hand, resting on the palm, and pressed by the thumb on one side, and the extremities of the fingers on the other. A light, but smartly repeated rap or two on the top of the box with the knuckles of the right hand constituted the commencement of the fourth operation, which ended by taking hold of the upper portion of the box with the fingers of the right hand, in the same manner that the lower was held by the left, and gently raising it obliquely, as it were, upon a hinge, although it possessed none, and leaving it, when nearly perpendicular, in charge of the now disengaged fore-finger and thumb of the left hand, whilst the right hand

---

* Why are ladies who indulge in this habit universally described as *elderly* ladies?

was entirely free. How radiant was the smile when the yellow dust filled at least a moiety of the cavity of the opened box. How disconsolate the expression when this devout consummation was not attained. Witness next the extended fingers, and the adroit dexterity with which the finger and thumb collected its accustomed dole, and conveyed it to the olfactory organs. How carefully it was carried, first to the right nostril, and then to the left, and with two hearty inspirations imbibed. The returning fingers now closed the box, which received another wipe, and was then returned into the receptacle. The fingers first, and then the nose, underwent the same purifying process by means of the brown handkerchief. Then, although no particle of dust could anywhere be seen, the whole frontispiece, from the chin to the knees, underwent a regular dusting; the handkerchief was replaced among the folds of the dress, the apron smoothed down with both hands, a half-uttered exclamation of satisfaction, and the work which had been temporarily laid aside was now resumed, until another occasion of a like character should arise to demand its suspension.

But to return to coca, the effects of which are described as of the most extraordinary nature, totally distinct from those produced by any other known plant in any part of the world. The exciting principle is said to be so volatile, that leaves, after being kept for twelve months, entirely lose their power, and are good for nothing.

Large heaps of the freshly-dried leaves, particularly while the warm rays of the sun are upon them, diffuse a very strong smell, resembling that of hay in which there is a quantity of melilot. The natives never permit strangers to sleep near them, as they would suffer violent headaches in consequence. When kept in small portions, and after a few months, the coca loses its scent, and becomes weak in proportion. The novice thinks that the grassy smell and fresh hue are as perceptible in the old state as when new. Without the use of lime, which always excoriates the mouth of a stranger, the natives declare that coca has not its true taste, a flavour which can only be detected after long use. It then tinges green the carefully swallowed saliva, and yields an infusion of the same colour. Of this infusion Pöppig made trial, and found that it had a flat, grass-like taste, but he experienced the full power of its stimulating principles. When taken in the evening, it was followed by great restlessness, loss of sleep, and generally uncomfortable sensations, while from its exhibition in the morning, a similar effect, though to a slight degree arose, accompanied with loss of appetite. Dr. Archibald Smith of Huanaco, when on one occasion unprovided with Chinese tea, made a trial of the coca as a substitute for it, but experienced such distressing sensations of nervous excitement, that he never ventured to use it again. It is not at all uncommonly used in this way; and the Indians have tea-parties or *tertulias*, for taking the infusion of the leaves, as well as for chewing them. Some affirm that in the coca-tea drinkings the effects are agreeably exhilarating. It is usual to say on such occasions, "*Vamos à coquear y acullicar*"—"Let us indulge in coca."

Chewing the coca becomes quite a passion in those who indulge in it; and when the habit is once commenced, it is affirmed that it is never discontinued, and that an instance of a reclaimed coquero has never been known. To indulge in the enjoyment of this narcotic, the Peruvian will expose himself to the greatest dangers. As its stimulus is most fully developed when the body is exhausted with toil, or the mind with conversation,

> the victim then hastens to some retreat in a gloomy native wood, and flinging himself under a tree, remains stretched out there, heedless of night or of storms, unprotected by covering or by fire, unconscious of the floods of rain, and of the tremendous winds which sweep the forest, and after yielding himself for two or three entire days to the occupation of chewing coca, returns home to his abode, with trembling limbs, and a pallid countenance, the miserable spectacle of unnatural enjoyment. Whoever accidentally meets the coquero under such circumstances, and by speaking interrupts the effects of this intoxication, is sure to draw upon himself the hatred of the half-maddened creature. The man who is once seized with the passion for this practice, if placed in circumstances which favours its indulgence, is a ruined being. Many instances were related to Pöpping while in Peru, where young people of the best families, by occasionally visiting the forests, had begun using the coca for the sake of passing the time away, and acquiring a relish for it, from that period been lost to civilization; as if seized by some malevolent instinct, they refused to return to their homes, and resisting the entreaties of their friends, who occasionally discovered the haunts of these unhappy fugitives, either retired to some distant solitude, or took the first opportunity of escaping, when they had been brought back to the towns.

So seductive becomes this habit, for we cannot doubt the veracity of these statements, that neither home, nor friends, nor family, nor society, nor fear, nor love, nor respect, nor any other creature, nor passion, would seem to have the power of winning them back from their monomania to a rational state of existence.

The virtues of the coca must be of the most astonishing character. The Indians, who are addicted to its use, are declared to be thereby enabled to withstand the toil of the mines amidst noxious metallic exhalations without rest, food, or protection from the climate. They run hundreds of leagues over deserts, and plains, and craggy mountains, sustained only by the coca and a little parched corn; and often too, acting as mules in bearing loads through passes where animals cannot go. Some have attributed this frugality and power of endurance to the effects of habit, and not to the use of coca; but the Indian is naturally voracious, and it is known that many Spaniards were unable to perform the Herculean tasks of the Peruvians until they habitually used the coca; moreover, it is affirmed, that without it, the Indians lose both their vigour and powers of endurance. During the siege of La Paz in 1781, when the Spaniards were constantly on the watch, and destitute of provisions, in the inclemencies of winter, they were saved, as chroniclers narrate, from disease and death by resorting to this plant. Some of those who deny many of the effects, said to be produced by its use, admit that the coca is useful medicinally as a preservative against the fevers which are consequent to a climate like that of Peru.

Hallucinations result from the use of the coca as from that of the narcotic hemp, but not, as it would appear, to the same extent. The inordinate use of this plant, as indeed of all the narcotics, seems to be attended with fearful results. One description with which we are acquainted, gives details of no very desirable character. It affirms that the abuse speedily occasions bodily disease, and detriment to the moral powers, but that still the custom may be perserved in for many years, especially if frequently intermitted, and the coquero sometimes attains the age of fifty with comparatively few complaints. But the oftener the orgies are celebrated, especially in a warm and moist climate, the sooner are their destructive effects made evident. For this reason, the natives of the cold and dry districts of the Andes are more addicted to the consumption of coca than those of the close forests, where undoubtedly other stimulants do but take its place. Weakness in the digestive organs, which, like most incurable complaints, increases continually in a greater or less degree, first attacks the unfortunate coquero. This complaint, which is called "opilacion," may be trifling at the beginning, but soon attains an alarming height. Then come bilious obstructions, attended with all those thousand painful symptoms which are so much aggravated by a tropical climate, jaundice and derangement of the nervous system follow, along with pains in the head, and such a prostration of strength, that the patient speedily loses all appetite. The whites of the eyes assume a leaden colour, and a total inability to sleep ensues, which aggravates the mental depression of the unhappy individual, who, spite of all his ills, cannot relinquish the use of the herb, to which he owes his suffering, but craves brandy in addition. The appetite becomes quite irregular, sometimes failing altogether, and sometimes assuming a wolfish voracity, especially for animal food. Thus do years of misery drag on, succeeded at length by a painful death.

This property of dispelling sleep, as a result of the inordinate use of coca, was noticed by Weddell, as the result also of the moderate indulgence, by way of experiment, in an infusion of the leaves, and which led him to suppose that the chemical principle of tea, called theine, would be found present in them. Professor Frémy analyzed them accordingly, but found no such principle was found, peculiar to this plant, the full properties of which are still unascertained.

Coca has the reputed power of sustaining strength in the absence of any other nutriment. The Indians declare, that when using it they feel neither the pains of hunger nor of thirst, that they are enabled to perform the most laborious operations with little or no food, insensible either to cold or weariness; that by its use they can ascend the steep passes of the Andes, carrying with them heavy loads, and without lassitude or loss of breath. When Tschuddi was in the Puna, he drank always before going out to hunt, a strong infusion of coca-leaves. Then, he states, he could during the whole day climb the heights, and follow the wild animals without experiencing any greater difficulty of breathing than he would have felt in similar movements along the coast. One account states, that a native, who was employed in laborious digging for five days and nights, tasted no food during that period, and only slept two hours each night. He regularly chewed the coca-leaves, to the extent of about half an ounce every two or three hours, and kept a quid of them constantly in his mouth. The work being finished, he went a two days' journey of twenty-three leagues across the level heights, keeping pace with a mule, and only halting to re-plenish his quid. At the end of all this labour, he was willing to engage for the performance of as much more without food, but with a plenti-ful allowance of coca. This man was sixty-two years of age, and was never known to have been ill in his life. For this reason, that it ap-pears to act as a substitute for food, several earned and ingenious au-thors have lamented that it has not been introduced into countries like our own, where it would be a boon so valuable to the poor in times of scarcity and distress.

What says science concerning this extraordinary power? One of two things is certain: either that the coca contains some nutritive principle which directly sustains the strength, or it does not contain it, and, therefore, simply deceives hunger while acting on the system as an excitement. As to the existence of a nutritive principle in coca, although it cannot positively be denied, on account of the quantity of nitrogen, together with assimilable carbonized products, which have been found to exist in the leaf; yet their proportion is so small com-pared with the mass, and especially with the quantity that a coquero consumes at once, that they can scarcely be taken into consideration. Moreover, it has also been affirmed that coca, as it is habitually taken, does not satiate hunger. The Indians who accompany travellers, will chew the leaves during the day, but, on the arrival of evening, they will fill their stomachs like fasting men, devouring, at a single meal, enough to satisfy an ordinary man for two days. The Indian of the Cordillera is like the vulture of his mountains, when provisions abound, he gorges himself greedily, when they are scarce, his robust nature enables him to content himself with very little. This is the evidence—what is the verdict? That the use of the coca assists, perhaps, to sup-port the abstinence; but that its action is confined to an excitement of a peculiar kind, very different from that of the ordinary excitants, and especially alcohol. Brandy gives strength, but that strength is only a loan, at the expense of strength reserved for the future. The stimulus produced by coca is slow and sustained, in part owing to the manner of its employment, as the infusion acts differently from the leaf as taken in the ordinary way tea and coffee act specially on the brain, on which they produce an anti-soporific effect; but while coca produces a little of this effect when taken in large doses, it does not act perceptibly upon the brain in small doses. To account for the ordinary effects of the leaf, one must suppose that its action, instead of being localized, as in the case of tea and coffee, is diffused, and bears upon the nervous system generally, producing a sustained stimulus, calculated to impart to those under its influence, that support which has been attributed erroneously to peculiar nutritive properties.

Superstition and prejudice combined have, however, ennobled this plant in the mind of the Peruvian, and he looks upon it as a true "gift of God." Its influences and effects are magnified in his own mind into something miraculous, and, indeed, miraculous powers have been attributed to it, for in what other light can we regard the belief current amongst them, that if the miner throws the masticated leaves upon the hard and impenetrable veins of metal, the ore will thereby become softened and be more easily worked? or that the leaves when placed in the mouth of a dead person, ensures it a more favourable reception into the world of spirits? or that when a mummy is met with disentombed from its narrow home, the presentation of a few leaves propitiates its disengaged spirit, and is accepted as a pious offering?

Much of the fidelity of the Indian to his coca, as with the smoker to his pipe of tobacco, is due to habit, and in this case the influence of the habit is more powerful, inasmuch as it has been handed down through a long line of ancestors, and is almost the only one which has been preserved. Finally, he finds in its use a distraction, and the only one, which breaks the monotony of his existence. The Peruvian Indians are of a gloomy temperament, and subject to fits of melancholy. When not engaged in out-door work, they will sit in their huts chewing coca and brooding gloomily over their own thoughts; indeed, the combined testimony of travellers establish the fact, that there is in their features an expression of concentrated melancholy, which seems to speak of an undefined but constant suffering; we cannot be astonished at finding such people seeking for comfort in the best substitute for opium that their country will furnish.

Coca appears to enjoy an undisputed reign in the Cordilleras; no other narcotic starts up to share the throne, and this is almost the only one which has not been imitated, or for which some substitute has not either been proposed or used. The antipodes, or nearly so, of this county possesses a plant, which, had it grown freely in other parts of the world might have been heard of more extensively as an indulgence. In Siberia, however, there seems to be little use made of the small indigenous rhododendron, which claims to be one of the most powerful narcotics in the world. Steller, the Russian botanist, had a tame deer which became so intoxicated by browsing on about ten of its leaves, that, after staggering about for some time, it dropped into a deep but troubled sleep for four hours, after which it awoke, apparently free from pain, but would never touch the leaves again. Steller's servants, after this, took to intoxicating themselves with the leaves without any evil effects. We have also been informed that certain of the Russians have been charged with the habit of following the example of these experimentalists, by getting drunk upon the leaves, which have been used in infusion, as Pallas states, with good effect in the cure of chronic rheumatism. The flowers of another species of rhododendron are eaten as a narcotic by the Hill people of India, but in these instances the extent of their use is so small, and the persons indulging in them so few, that no claim can be set up for them, except as minor narcotics occasionally employed, when the other and more important substances cannot readily be obtained.

For the basis of much which this chapter contains, we are indebted to the travels in Bolivia and Peru of that worthy trio of doctors, Pöppig, Weddell, and Tschuddi, besides three times as many more, less noted and less known, but whose information was not less to be relied upon on the points concerning which they have spoken. Whether the votaries of our Lady of Yongas are as numerous as has been asserted, or only of the number we have suggested—whether the influence of this plant over the stomachic regions is sufficient to subdue the pangs of hunger, or allay the cruelties of thirst, or these are only effects due to the imagination—whether it has the marvellous power of softening the adamantine rock or strengthening and supporting the lungs in the ascent of Andean summits, or whether these, and all of these, are fictions proceeding from the heat-oppressed brain, it is, nevertheless, certain, that a great amount of interest gathers around this plant, which associates itself so intimately with the country in which it flourishes, that, as for centuries past, so for centuries to come, coca will remain the characteristic plant of the Peruvian nation, as tea was, and is, of the Chinese.

CHAPTER XXII

WHITEWASH AND CLAY

Alexander died. Alexander was buried. Alexander returneth
into dust; the dust is earth; of earth we make loam. And why
of that loam, whereto he was converted, might they not stop
a beer barrel?

Hamlet

The fact, at one time doubted, but now es-
tablished beyond dispute, that some tribes indulge in the habit of dirt-
eating, is one which, from its singularity, claims notice. The Malayan
uses lime as an ingredient in compounding his favourite masticatory,
and the coquero of the Andes mixes it with his leaves of coca. The
Nubians mingle the saline natron with their quid of tobacco, and the
blacks of Gesira the same material to compound their bucca. The
Ottamacs and Omaguas avail themselves of the assistance of shell lime
to give pungency to their intoxicating snuffs. The tribes on the coast
of Paria, according to Gomara, stimulated the organs of taste by caus-
tic lime, as other races employ tobacco, coca, or betel. In our own
days this practice exists among the Guajiros at the mouth of the Rio
de la Hacha. Here the still uncivilized Indians array small shells, cal-
cined and powdered, in a box made from the husk of a fruit. This box
is suspended from their girdle, and serves a variety of purposes. The
powder used by the Guajiros is an article of commerce, as formerly
was that of the Indians of Paria. What could first have induced these
people to use by itself, or other races to mingle with vegetable substances, a mineral only known to us as a whitewash, or for somewhat
similar vulgar uses, and to metamorphose it into a luxury, is difficult

to understand. We comprehend the value of lime when stirred about in a pail, with sufficiency of water to reduce it to the consistence of cream, and then by the aid of a broad flat brush transferred to the ceilings of our dwellings. We cannot so well comprehend or appreciate the luxury of rolling it into a pellet, and transferring it to our mouths, as a whitewash for regions where the curious eye of man does not penetrate.

The residents at the fur-posts on the Mackenzie River, have a mineral in use among them, known by the appellation of *white mud*, which is used for whitewashing, and when soap is scarce, it supplies the place of that article for washing clothes. It resembles pipe-clay, and exists in beds from six to twelve inches in thickness. It is of a yellowish white colour, sometimes with a reddish tinge. On the Arkansas also a similar substance has been met with, called *pink clay*. The clay of the Mackenzie is smooth, and, when masticated, has a flavour, we are told, resembling the kernel of a hazel nut. Sir John Richardson obtained some of this clay in his journey to Prince Rupert's Land, and had it examined, but could not discover in it any nutritious properties, or detect the remains of infusorial animalculæ, such as are found in other edible clays. The natives of the locality in which this substance is found, eat it in times of scarcity, and suppose that by its use they prolong their lives. There are certain physiological reasons known to us whereby we account for fowls, and other winged bipeds indulging in the singular propensity of swallowing small pebbles, fragments of lime or mortar, sand and clay; but as we cannot apply these same arguments to the cases of other "bipeds without feathers" who indulge in the same propensity, we naturally seek for some signs of nutritious value in the substance itself. In this instance the remote probability of its containing decayed animal matter does not apparently exist, for the microscope detects no infusoria. And unless we argue, as did Hamlet with his friend Horatio, that in this clay are the remains of a previous generation, we can scarce account for its being a good article of food.

> Imperial Cæsar, dead and turned to clay,
> Might stop a hole to keep the wind away;

or dead Indians turned to clay to appease the hunger of their living descendants. Thus, if the imagination may trace the noble dust of Alexander, till we find it stopping a bunghole, may it not also follow this same clay from the bunghole into the veins of a new Alexander?

Richardson states that the above is a kind of pipe-clay. If made into pipes for smoking, Hamlet might argue still further, "may we not trace the dust of the dead Indian, till we find a man smoking his weed from the leg or arm of his great grandfather."

Clay eating exists in South America, among the Guamos, and by the tribes between the Meta and the Apure. The natives here speak of the custom as one of great antiquity. The Ottomacs are, however, great clay-eaters. Humboldt found amongst them heaps of earth-balls, piled up in pyramids three or four feet high, and these balls five or six inches in diameter. This clay was of a yellowish grey colour, and did not contain magnesia, but silex and alumina, and three or four per cent. of lime, no trace of organic substance either oily or farinaceous, could be found mixed with it. If the Ottomac is asked what he lives upon during the two months of the inundation of the rivers, he shows you his balls of clayey earth. It is asserted that far from becoming lean at that season, they are, on the contrary, extremely robust.

At the village of Banco, on the Rio Magdalena, the same traveller found Indian women making pottery, who continually swallowed great pieces of clay.

On the coast of Guinea, the negroes eat a yellowish earth, which they call *caouac*, the taste of which is said to be agreeable, and to cause no inconvenience. When these Africans are carried to the West Indies, they still indulge in the custom, for which purpose Chanvalon states that it is sold in the markets, but that the West-Indian clay does not agree with them so well as that of their native country.

Labillardière saw between Surabaya and Samrang little square reddish cakes, called *tanaampo*, exposed for sale, which were slightly baked, and eaten with relish.

Leschenault states that the reddish clay (*ampo*) which the Javanese are fond of eating occasionally, is spread on a plate of iron and baked, after being rolled into little cylinders in the form of cinnamon bark. In this state it is sold in the markets. It has a peculiar taste, which is owning to the baking, is very absorbent, and adheres to the tongue. The Javanese women eat the *ampo* in order to grow thin, the absence of plumpness being there regarded as a kind of beauty.

In times of hunger or scarcity, the savages of New Caledonia eat great pieces of a friable stone, which contains magnesia and silex, with a little oxide of copper.

The African negroes of Bunck and Los Idoles eat a kind of white and friable steatite, or soap-stone, from which custom they are said to suffer no inconvenience.

At Popayan and several of the mountainous parts of Peru, finely-powdered lime is sold in the public markets with other articles of food. This powder is, however, generally mixed with the leaves of the coca, and used as a masticatory. In other parts of South America, lime is swallowed alone, the Indians carrying with them a little box of lime, as other people carry their tobacco-box, snuff-box, or siri-box.

In the kingdom of Quito, the Tigua natives eat from choice, and without any ill consequences, a very fine clay mixed with sand. This clay, mixed with water, renders it milky. Large vessels filled with this mixture, called *agua de llanka*, water of clay, or *leche de llanka*, milk of clay, may be seen in most of their huts, where it serves as a beverage.

On the banks of the river Kamen-da-Maslo, there is produced a fossil, or an earthy substance, called in Russian *kamennoye maslo*, stone butter, which is eaten in various ways, as well by the Russians as the Tongousi, it is of a yellowish cream colour, and not unpleasant in taste, but it is forbidden as pernicious in its effects. This earthy matter is stated to be a fossil, or salt oozing out of rocks, in many parts of Siberia, but chiefly from those near the river Irtish and Yenissei. When it is exposed to the air in dry weather it hardens, but in wet weather it again becomes soft or liquid. The Russian hunters use it also as a bait. The animals scent it from afar, and are fond of the smell.

In Germany, the workmen employed in the quarries of sandstone at Kiffhauser, spread a fine clay upon their bread instead of butter, which they call *steinbutter* (stone butter). There is another substance, called *bergbutter*, or mountain butter, which is a saline substance produced by the decomposition of aluminous schists.

On the shores of a lake near Urania, in Sweden, is found a deposit, called by the peasants "mountain meal" *(bergmehl)* which they use, mixed up with flour, as an article of food. This deposit consists chiefly of fossil infusoria.

In Finland also, a similar kind of earth is mixed with bread stuff, as also in parts of Northern Germany in cases of scarcity or necessity. In Lapland also, this fossil farina has been found, and applied to a like use. The Tripoli or rotten stone of commerce is an infusorial earth of this description, composed of fossils of extraordinary minute dimensions.

A poor man, in the neighbourhood of Dejufors, Sweden, some years since, found an earth of this description, which had much the appearance of meal. The people being at that time in a state of privation, and living upon bark bread, this man took some home, mixed it with rye meal, baked it into bread, and found it palatable, whereupon there was a general run upon this earth, and some of it found its way to Stockholm. On analysis it was found to contain flint and felspar, finely pulverized with lime, clay, oxide of iron, and some organic substance resembling animal matter, and yielding ammonia, and an oil.

Ehrenberg found that a hill in Bohemia was one mass of the siliceous fossil shells of these minute creatures, and that in a stratum fourteen feet in thickness, one cubic inch contained the remains of 41,000,000,000 of individuals.

These kind of deposits are continually accumulating, and producing important changes, in the bed of the Nile, at Dongola, and in the Elbe, at Cuxhaven, and even choking up some of the harbours in the Baltic Sea.

Dr. Trail analyzed a bergmehl from the North of Sweden, and found it to be composed of the minute shields of infusoria, about one thousandth of an inch in size, consisting chiefly of siliceous earth and alumina. A small quantity of this curious substance was found in County Down, Ireland, by Dr. Drummond, twenty years ago, while sinking a pit near Newcastle.

MM. Cloquet and Breschet ate experimentally as much as five ounces of silvery green laminar talc. Their hunger was completely satisfied, and they felt no inconvenience from the use of a kind of food to which they had not been accustomed. In parts of the East, use is still made of the Bole earths of Lemnos, which are clay mixed with oxide of iron.

In Portugal and Spain, *bucaro* clays are made into vessels, from which many are fond of drinking on account of the smell of clay; and the women of the province of Alentejo acquire a habit of masticating the bucaro earth, and feel it a great privation when unable to indulge in this vitiated taste.

In the Bolivian markets, Dr. Weddell saw a grey-coloured clay which was offered for sale. It is called *pahsa*, and the Indians of La Paz eat it with the bitter potato of the country. It is steeped in water, made into a kind of gruel, and seasoned with salt.

At Chiquisaca a kind of earth called *chaco* is made into little pots, and eaten like chocolate. Although their moderate use is not calculated to injure the system, their contribution to the nourishment of the body must be but small.

In the valleys of the Sikkim Himalayas, a kind of red earth is chewed as a cure for the goître, but it is not stated to be regularly indulged in as an article of food either there or in any other part of India.

Mr. Wallace relates that a little Indian boy died from the habit of dirt-eating—a very common and destructive habit among Indians and half breeds in the houses of the whites in the Amazon Valley. All means had been tried to cure the lad of the habit. He had been physicked and whipped, and confined in doors; but when no other opportunity offered, he would find a plentiful supply in the mud walls of the house. The whole body, face, and limbs swelled, so that he could with difficulty walk, and not having so much care taken of him, he ate his fill and died.

Those who have had much to do with children, will have noticed amongst some of them the germs of this propensity, which will occasionally develop itself in chewing pieces of pipe, slate pencil, chalk, and other substances of a like nature. Although not carried to so great an extent as to become injurious, cases of this kind are far from being, among school children, either exceptional or uncommon.

In the mission of San Borja, Humboldt found the child of an Indian woman, which, according to the statement of its mother, would hardly eat anything but earth. It was very thin and emaciated.

These instances are not, after all, so singular as those of habitual, national dirt-eating which we find amongst the tribes of South America and the negroes of Africa. Children are not always the most particular in the choice of their articles of food, or we should not read of such instances as occur in tropical America of these youngsters drawing immense centipedes out of their holes and eating them; or, as related by Captain Cochrane, of a child devouring several pieces of tallow candle, which was succeeded by a large lump of yellow soap, all of which he seemed to enjoy.

Chroniclers often make mention of the employment, during times of war, of kinds of infusorial earth as food, under the general term of mountain meal. This was the case in the Thirty Years War, at Camin in Pomerania, Muskau in the Lausitz, and Kleiken in the Dessau territory; and subsequently in 1719 and 1733 at the fortress of Wittenberg. But in times of war and scarcity, one is prepared to hear of men satisfying their hunger by every legitimate means.

M. S. Julien sent to the Academy of Sciences at Paris some few years since, specimens of a peculiar mineral substance from the province of Kiang-si in China, on which, in times of famine, the inhabitants have been said to be able to support themselves as a nutriment. It has a disagreeable taste, and produces dryness in the mouth. It is nevertheless used by the natives mixed with flour, and is even esteemed by them.

It may appear somewhat singular to refer to these dirt-eating customs, in connection with those relating to narcotics. The connection is, however, more intimate than at first glance might appear. Two kinds of substances are mostly resorted to, either to gratify these depraved tastes, or satisfy the cravings of hunger—lime and clay, or, as we have designated them—*clay* and *whitewash*. It is, or has been matter of dispute, whether the stimulating properties of the betel and coca, and the intoxicating snuffs of the Orinoco, are to be attributed to the vegetable substances themselves, or to the lime used with them, or both in conjunction; hence the introduction of lime is not considered appropriate. As for the clay, it is not only intimately associated with the other, from the similarity of the use to which it is thus strangely applied, but the connection of it in some of its forms with the consumption of one or two of the narcotics, as the means whereby they are indulged in, must serve as an apology, if such be needed.

CHAPTER XXIII

PRECIOUS METALS

*The virtues of the noble metals are, moreover, of such a
nature that they inspire respect even in those who do not seek
these qualities in higher spheres, but ask after the common
and every-day usefulness of a thing.*

Von Kobell

Some consider those metals most pre-
cious which, like gold and silver, have earned that reputation by acting in the capacity of representative of wealth, as the current coins of
civilized nations. To some men these have been esteemed more precious than health, or even than life itself; others, calculating on the
grounds of utility, have considered iron and copper, so universally applicable to the wants of civilized life, such mighty agents in the cause
of civilization, as the most precious of metals; and these may be right
in their calculations, for although we might manage to get on without
the former, we can hardly imagine for ourselves the condition occasioned by the loss of the latter. There are yet a few to whom it would
seem, however strange the fact may appear, that two metals are the
most precious which the rest of the world have no idea of considering
as of but a very low rate of value, and without which they can readily
conceive of the world moving on without any very great sense of their
loss. These two are arsenic and mercury. The very names are almost
sufficient to send a shudder of horror through us as we write or repeat
them; and to elect them into the highest place in our affections is the
last act we should, in a state of sanity, deem ourselves likely to perform. The one suggests images of Aqua Tophana and the Middle Ages,

and our teeth loosen in our gums with unpleasant reminiscences of black draught and blue pill as associated with the other. For one we can think of no better employment than the extirpation of rats, or the preservation of mummies; and for the other no more exalted an occupation than to coat the backs of our mirrors, or inform us of the conditions of the atmosphere. That any one could indulge in them as luxuries, or, by their habitual use, elevate them to a companionship with tobacco and opium, with haschisch and coca, would appear to be a gross libel upon the *Seven Sisters of Sleep*, and a satire upon the cherished companions of millions of the human race.

Medical men, foremost amongst whom is Dr. Christison, consider that these minerals cannot be indulged in without exercising a deleterious effect upon the system. The cumulative action of mineral poisons is a great point of difference between them and those of vegetable origin, for although the same eminent physician is of opinion that tobacco may be indulged in without injury, he does not believe such a possibility to exist with regard to mercury and arsenic.*

The use of corrosive sublimate, the bichloride of mercury, is certainly restricted within very confined limits, and even within those limits, the information we have is very meagre. At Constantinople, the opium-eater, who finds his daily dose insufficient in time to produce those results which at first accrued from its use, resorts to the expedient of mixing therewith a small quantity of corrosive sublimate, to increase the potency of the drug. By itself, it is never indulged in as a passion in the same manner as vegetable narcotics, nor can the same pleas be urged in favour of its use, or in extenuation of its abuse. An opium-eater at Broussa is stated to have been accustomed to swallow daily with his opium, forty grains of corrosive sublimate without any apparently injurious effects. In South America its use is affirmed to be very extensive.

Arsenious acid, or white arsenic, is a more popular irritant than mercury. The arsenic-eaters of Styria are now historical individuals, and the custom there and in the neighbouring districts appears to be a common one among the labouring population. Itinerant pedlars vend it for this purpose, and it becomes a necessary of life to those who

---

*This name, derived from the Greek, indicates *strong, powerful*.

commence the practice. It is taken every morning as regularly as the Turk consumes his opium.

One of the benefits said to accrue from its use is, that it gives a plumpness to the figure, softness to the skin, freshness to the complexion, and brilliancy to the eye. For this purpose, young men and maidens resort to it, to increase their charms, and render themselves acceptable and fascinating to each other. A friend, recently returned from Canada and the United States, informs us, positively, that it is largely consumed by the young ladies, in those regions of the civilized world, for the same purposes above described, to which it is resorted by the Austrian damsels. He declares that the custom is so common that no surprise is excited on discovering any one addicted to its use, and that amongst the fairer sex it is the rule rather than the exception.

The principal authority for its use in the European districts, is the celebrated traveller Von Tschuddi, who has published an account of several cases which have come to his knowledge. In one instance, a pale, thin damsel, anxious to attach herself to her lover, by presenting a more prepossessing exterior, took the "precious metal," in the form of its oxide, several times a week, and soon became stout, rosy, and captivating; but in her over-anxiety to heighten her charms, and rival the fabled beauties of old, and having experienced the benefit of small doses of the poison, ventured upon a larger quantity, and died from its effects, the victim of her vanity. The habit is generally commenced with small doses, starting with about half a grain or less, each day, and gradually increasing it to two or three grains. The case of a hale old peasant is mentioned, whose morning whet of arsenic reached the incredible quantity of four grains.

Another singular benefit is supposed to arise from the use of this substance, similar to that claimed by the Peruvians for their coca, namely, that of rendering the breathing easier in toiling uphill, so that steep heights may be climbed without difficulty or exhaustion. It is curious that the mountaineers of the Andes and the Alps, at distances so remote, should deem themselves possessed of the means of assisting nature in surmounting difficulties, by preventing exhaustion in climbing the mountain side: in one instance, by chewing a quid of leaves which grow plentifully on the mountain slopes, and in the other, by swallowing a small fragment of a mineral obtained from the mines at the mountain side.

Whilst the practice of arsenic eating is continued, no evil effects would seem to be experienced, everything connected with the body of the eater seems to be in a flourishing condition, the appearance is healthy, plump, and fresh, no symptoms of poisoning are manifested until the regular dose is discontinued, when a great feeling of discomfort arises, the digestion becomes deranged, burning sensations and spasms are present in the throat, pains in the bowels commence, and the breathing becomes oppressed. From these unpleasant sensations there is no relief but by an immediate return to the habit of arsenic-eating, and hence, when once commenced, the use of this article becomes a necessity of life, and the poisonous mineral a "precious metal."

Dr. Macgowan of Ningpo, says,

> We are told that Mongolian hunters, beyond the wall, eat arsenic to enable them to endure cold when patiently lying on the snow to entrap martins. In this part of China arsenic is taken by divers, who in cold weather plunge into still water in pursuit of fish, which are then found hybernating among stones at the piers of bridges. We perceive with regret, that the modern Chinese have added arsenic to their habitual stimulants. The red sulphuret in powder is mixed with tobacco, and their joint fumes are smoked in the ordinary manner. We have met with no habitual smokers of this compound of mineral and vegetable poisons; but persons who have made trial state that dizziness and sickness attend first attempts. After a few trials, arseniated tobacco may be taken without any apparent inconvenience. From reports given of it, we infer that its effects on the Chinese are analoguous to what is observed among the arsenic-eating peasants of Austria.
>
> At Peking, where arseniated tobacco is most in use, it costs no more than the unmixed article; it may be known by the red colour imparted to the vegetable by the powdered proto-sulphuret. Its introduction is attributed to Cantonese from Chauchau. If this be correct, it is probable that these southerners, unable at the north to procure the masticatory to which they are addicted, sought to appease a craving for the pungent but harmless lime and betel nut, by substituting the deleterious mineral gas. Many of the miserable victims of opium, to whom the narcotic is a necessity, and not a pleasure, have eagerly employed the new stimulant to prop and exhilarate their exhausted

bodies, and, perhaps, have thereby meliorated and prolonged their existence. We would fain hope that the use of arsenical stimulants will not become general; yet that pernicious custom is extending, and we know our race too well not to entertain fears on that subject. It is even stated that, for a time at least, the reigning Emperor in his boy-hood preferred tobacco thus mineralized. In domestic economy, the red sulphuret is employed for making away with rats and husbands.*

One of the best things that Hahnemann ever did was to write a treatise on arsenic. This he did well, and therefore deserves to be re-membered; but for this he is often forgotten, and is only extolled for a less important labour—the introduction of homœopathy. Chemists deserve well of mankind for the assiduity with which they have stud-ied this subtle poison, so that now it may be detected in the minutest quantities. One point, however, seems to be hardly clear, and on this, perhaps, the Styrian peasant could enlighten us, namely, the taste of arsenic, some declaring that it has no distinguishable taste, others, that it is sweetish, and others saline. The only means of arriving at the truth is rather too hazardous a one to be ventured upon.

The effects of arsenic upon the human frame, were illustrated in a curious case which occurred a few years since in the northern part of France. A domestic at a country seat wished to cause the death of his mistress, and mixed arsenic in small quantities with her food, hoping that the slow operation of the poison would prevent any suspicion of murder. To his great astonishment, she gained rapidly in health, flesh, and spirits. At length he gave her a larger quantity, which occasioned serious illness, and led to the discovery and punishment of the crime.

We have as yet applied arsenic only to some of the purposes for which it is applicable. The roses of England possess enough of bloom without resorting to the bloom of the smelting furnace. Although we use it to preserve with all the appearances of life the deceased zoologi-cal curiosities of our museums, we do not seek its aid to enhance the charms of those living specimens of beauty which are the glory and pride of our hearths and homes. Fortunately, we have no Andes to climb, and no Alps to scale, and the summits we have to gain are arrived at by dint of perseverance, and no small amount of puffing, in

---

*Edinburgh Medical Journal, 1857.

which latter circumstance it seems to be our nature to glory as much as the Peruvian or the Austrian in its absence. Now and then we become suspicious of its presence in our green paper hangings, and in that menial office are almost content to dispense with its services. Or anon, we are treated to a scramble of Bradford drops, which, finding the temperature of the climate uncongenial, melt away to a stray ghost or two that haunt the stoppered bottles of our chemical museums. Grumble as we may at *our* precious metals, we—

> *Rather bear those ills we have,*
> *Than fly to others that we know not of.*

Animals have not escaped arsenic-eating, for the Austrians, having discovered its property of plumping up, and putting into good condition the human animal, have resorted to it, as an improver of their ill-conditioned horses. Gentlemen's grooms bestow it upon the animals in their charge, and pronounce its effects as certain and as marvellous, as upon thin and sickly-looking damsels. A pinch of the white powder is sprinkled like pepper over the "feed of corn," or tied up in a piece of rag and fastened to the "bit," before that instrument is introduced into the animal's mouth. The same two properties are said to be exhibited in the case of the horses, as are affirmed to take place in man. The body is plumped out, and rounded into fair proportions, the skin rendered sleek and glossy, and the breath is improved, so that long journeys, steep and rugged ascents, and heavy loads, are readily overcome by its potency. If this secret were communicated to some of our London omnibus and cabmen, it would probably be of advantage to the appearance of some of the poor animals doomed for a certain time to *walk* this earth, and increase their facility for moving through a space of three or four miles in less time than a pedestrian could accomplish the feat.

The teamsters in mountainous countries, frequently add a dose of arsenic to the fodder, which they give their horses, before a laborious ascent. The practice of giving arsenic to horses may continue for years without accident, but as soon as the animal passes into the hands of a master who does not use arsenic, he becomes thin, loses his spirits, and, in spite of the most abundant nourishment, never recovers his former appearance.

The use of arsenic for horned cattle is less frequent; it is only given to oxen and calves intended for fattening. In Austria, hogs and other animals are also fattened by a careful use of arsenic.

Precious metals, like precious stones, are subject to misfortunes. As of the latter, a learned professor saith,

> Patents of nobility are distributed here in the most arbitrary manner, and outward aspect and character, weigh heaviest in the scales by which they are determined. To such an extent is this the case, that the stones which have literally and truly fallen from the skies, are not reckoned among the precious stones, although they have been in all times objects of curiosity to the most cultivated minds, and certainly are of *very high descent*, since they came, at least, from the moon, and are even imagined to be young worlds, little princes, which would in time have come to reign as planets. And whence this injustice? Because these little strangers, which, perhaps are pleased to travel *incognito*, have an inconspicuous exterior, are enveloped in a dark weather-proof cloak, because from under this cloak, only a greyish suit, without gold lace, with merely a little iron scattered about it, comes to light; because this aspect does not show from afar off that they have fallen from the skies, and because they do not say to everybody, "My mother lives in the mountains of the moon."

And although mercury, not only in name, but also in its volatile and skyward tendencies, claims kindred with the planetary system, which tendencies are likewise shown in the behaviour of the other metallic substance, of which this chapter discourses. Yet their *high* claims are disregarded, and, like the aerolites, they are condemned by the majority of men to a plebeian rank and menial offices.

CHAPTER XXIV

DATURA AND C♀.

That skulk in the depths of the measureless wood
'Mid the Dark's creeping whispers that curdle the blood.
Where the wolf howls aloof, and the wavering glare
Flashes out from the blackness the eyes of the bear.

The thorn-apple and nightshade are types of a class of narcotics, which, though not largely employed either for their intoxicating effects or their medicinal virtues, are, notwithstanding, extremely powerful in their effects, and, when used, exercise a wonderful influence upon the brain. The majority of them belong to that family of plants, of which, not only tobacco, but the potato, are members; so that, if only from their family connections, independently of any other right, they have a claim upon our attention and respect. Beyond this, even, we shall find them insinuating themselves into the good graces of that portion of the creation who have taken the two members of the family already named under its protection, and adopted them as companions, the one to soothe and console after the hours of labour are past, the other to aid in giving strength to perform that labour, or satisfy the cravings of hunger.

The solanaceous plants have, in general, narcotic qualities. In some species these are developed in a great degree, so as to render them extremely poisonous; in others, they are obscured by the prevalence of starchy matter. In some instances parts of the plant have narcotic properties, whilst other parts are used as articles of food. The bitter sweet (*Solanum dulcamara*) has slightly narcotic properties, and the

scarlet berries are considered poisonous. The common nightshade
(*Solanum nigrum*) has more active narcotic properties. The potato
(*Solanum tuberosum*) has slight narcotic qualities in its leaves and fruit,
but its tubers are edible and nutritious. The deadly nightshade (*Atropa belladonna*) is a highly poisonous plant, narcotic in all its parts.
Henbane (*Hyoscyamus niger*) contains also similar properties. Many
species of thorn-apple are powerfully narcotic, especially the seeds or
fruit; this is especially the case with our common thorn-apple (*Datura
stramonium*), with the thorn-apple of the Andes (*Datura sanguinea*),
and of North America (*Datura tatula*), the thorn-apples of India
(*Datura metel*, *D. ferox*, and *D. fatuosa*). Several species of *Nicotiana*
furnish tobacco. The fruit of different species and varieties of *Capsi-
cum*, which are used as pepper, possess irritant properties which obscure the narcotic action. Other species are used as narcotics, or as
poisons, and some, as the tomato and other Lycopersicums, as articles
of food; but the majority give evidence, in some of their parts, of the
existence of a narcotic principle.*

The Kala dhatoora (*Datura fatuosa*) and Sada dhatoora (*Datura
alba*) are very common species of thorn-apple over the peninsula of
India, where they are also called *mazil* or *methel*. For the purpose of
facilitating theft and other criminal designs, the seeds are very commonly given in Bengal, with sweetmeats, to stupify merely, but not
with the intention of killing. Intoxication or delirium is seldom produced. The individual sinks into a profound lethargy, with dilated
pupils, but natural respiration. These symptoms have been known to
continue for two days. The vision often becomes obscured long after
the general recovery takes place. Graham says that the seeds are often
fatally used for these purposes in Bombay. The narcotic action is more
speedy and powerful on an empty stomach than after a meal; hence

---

*The potato, the tomato, and egg plant possess, when uncooked, in a mild
degree, the properties of the nightshade, the stramonium, and the henbane,
confirming the remark of De Candolle "that all our aliments contain a small
proportion of an exciting principle, which, should it occur in a much greater
quantity, might become injurious, but which is necessary as a natural condi-
ment." In fact, when food does not contain some stimulating principle, we

death often ensues from the effects when the intention was only to produce narcotism.

In some parts of South America, especially in Peru, where a species of thorn-apple (*Datura sanguinea*) grows wild, the natives, in certain cases, drink a decoction of the leaves or seeds, which produces such violent effects as to cause them to fall into a state nearly resembling death, and lasting frequently two or three days. Every malady is there ascribed to enchantment, and this very singular plan is resorted to to discover by whom the mischief may have been wrought. In cases of extreme illness the decoction is given, not to the sick person, but to the nearest relative, who devotes himself for this purpose, to discover during his sleep the sorcerer or Mohari who has inflicted the disease. The medicine soon causes the relative to fall under its influence, and he is placed in a fit position to prevent suffocation. On returning to his senses he describes the sorcerer he has seen in his dreams, and the whole family sets out to discover the Mohari who bears the nearest resemblance to the description, who, when found, they compel to undertake the cure of the sick person. When no sorcerer has been seen in the vision, or no one is found resembling the one which has been seen, the first Mohari they meet with is obliged to undertake the office of physician. Should the patient die during the vision of the relative, the sorcerer whose image is then supposed to be presented is subjected to the same fate.

This plant, which is called "Florispondio" in tropical America, appears always to have played, and still continues to play, a prominent part in the superstitions of the natives. The Indians of Darien, as well as those of Choco, according to Seemann, prepare from its seeds a decoction, which is given to their children to produce a state of excitement, in which they are supposed to possess the power of discovering gold. In any place where the unhappy patients happen to fall down, digging is commenced; and as the soil nearly everywhere abounds with gold dust, an amount of more or less value is obtained. In order to counteract the bad effects of the poison, some sour *chica*, a beer made of Indian corn, is administered.

It is this same thorn-apple which is used amongst the Andes of New Granada, and even as far south as Peru, for the purpose of preparing therefrom a drink, with very strong narcotic properties, which

they call "Tonga." Dr. Von Tschuddi has given a description of the
effects of this narcotic upon an old Indian.

> Shortly after swallowing the beverage he fell into a heavy stupor.
> He sat with his eyes vacantly fixed on the ground, his mouth con-
> vulsively closed, and his nostrils dilated. In the course of about a
> quarter of an hour his eyes began to roll, foam issued from his half-
> opened lips, and his whole body was agitated by frightful convul-
> sions. These violent symptoms having subsided, a profound sleep of
> several hours succeeded. In the evening, when I saw him again, he
> was relating to a circle of attentive listeners the particulars of his
> vision, during which he alleged he had held communication with
> the spirits of his forefathers. He appeared very weak and exhausted.

By means of this plant they believe that they can hold communication with their ancestors, and obtain a clue to the treasures con-
cealed in their *huacas* or graves—hence it is called huaca-cacha or
grave plant. It has been supposed that the frenzied ravings, called
prophecies, of the Delphic oracles were produced by this plant, which
has been used, as Dr. Lindley asserts, in the temple of the sun at
Sogamossa, near Bogota, in New Granada, for the same purpose. Already we have alluded to the Delphic oracles more fully, when writing
of the "Sisters of Old."

The cunning few acquainted with some of the extraordinary properties of certain plants, which were unknown to the superstitious and
barbarous multitude in days gone by, had ample means at their dis-
posal for imposing on their credulity, by the performance of wonderful
cures, working apparent miracles, and gulling the less informed into
the belief that they were either in direct communication with the
spiritual world, or had received a divine commission by which to govern. Most of the marvels of ancient times were no greater than the
little experiments which the school-boy now performs for his amusement and that of his companions, with a few crystals and powders,
contained in as many pill-boxes.

The pots or gourds, in which cocoa-nut sap to make arrack is drawn
off in Ceylon, are sometimes visited and the contents carried off during
the night. To detect the thief, the leaves of a species of datura, or thorn-
apple are occasionally put into some of the pots. By means of the highly
intoxicating effect of this compound the marauder is often discovered.

On the Coromandel coast the retailers of toddy sometimes rub the inside of the pots with the seed-vessel or leaves of this highly poisonous plant, to increase the intoxicating influence of the toddy.

The phrase "pariah-arrack" is often used to designate a spirit distilled in the peninsula of India, which is said to be rendered unwholesome by an admixture of gunja, and a species of Datura, with the intention of increasing its intoxicating quality. It is not clear whether the term pariah-arrack be colloquially employed to designate an inferior spirit or an adulterated compound. It is curious that a system of "doctoring" beverages, to make them heady, should obtain abroad, as it does at home, and in both cases perhaps independently: for it does not seem probable either that we borrowed the system from the Hindoos, or that they copied it from us.

While under the influence of these narcotics the mind seems to be subjected to a troubled dream, and the person suffering from it indulges in fits of uncontrollable laughter. Beverley, the historian of Jamaica, quaintly describes the effects of the thorn-apple. Some soldiers, who were sent to quell the rebellion in the island, ate of it:

> The effect was a very pleasant comedy, for they turned natural fools upon it for several days. One would blow up a feather in the air, another would dart straws at it with much fury. Another, stark naked, was sitting up in a corner grinning like a monkey, and making mouths at them. A fourth would fondly kiss and paw his companions, and sneer in their faces with a countenance more antic than a Dutch doll. In this frantic condition they were confined, lest in their folly they should destroy themselves. A thousand simple tricks they played; and, after eleven days, returning to themselves again, not remembering anything that had occurred.

The extract of stramonium or common thorn-apple has occasionally, when injudiciously administered, produced similar effects upon the individual to whom it has been given, affecting the senses, particularly that of sight. "Imaginary objects are seen to play before the eyes, at which the victim strikes, as they seem to terrify him. And similar results have occurred from the use of the seeds." Fowler relates a case of a child who supposed that cats, dogs, and rabbits were running along the tops and sides of the room. Dr. Winslow says "that when inhaled, the smoke conveys a sense of gentle tranquillity, the muscles of the thorax, and those which have been called into action to assist them, in the paroxysms of asthma which the smoking is resorted to to relieve, are rendered less irritable and the fibre is relaxed, sleep is induced, but there is rarely any disturbance of the imagination."

In France and Germany, this plant has been resorted to for the basest of purposes, and many unhappy victims have been consigned to hopeless insanity by its means, details of which would be far more horrible than interesting. Faber also speaks of its use by the ladies of the Turkish harems; but there is doubt whether this is not one of those marvels, of which many may be met with in connection with medicinal agents, containing more of romance than reality. Dr. Ainslie states that the seeds form one of the ingredients of the confection of hemp and opium known under the name of *majoon* in India; as henbane is asserted to enter into the composition of that in use uder the same, or a similar name, in Egypt. The proportion of either of these when used is doubtless small, and is in most cases dispensed with.

Etymologists declare that the name of belladonna, which has been given to the deadly nightshade (*Atropa belladonna*), was so given because those to whom it was administered fancied they saw beautiful females before them.* There is no doubt that it produces illusions of a singular character, and cases of impulsive insanity have resulted from its use in repeated doses. The effect of belladona upon the brain is more extraordinary than those usually attendant upon the use of other narcotics. Persons who have been poisoned by the berries of the plant have become restless and delirious, complained of dimness of vision, and subsequently loss of sight. There were observed frequent spasmodic contractions of the muscles of the eyeballs and the throat, with strong symptoms of mania. Six soldiers who were poisoned by the plant exhibited delirium the most extravagant, and commonly of the most pleasing kind, accompanied with immoderate and uncontrollable paroxysms of laughter, sometimes with constant talking, but occasionally with complete loss of speech. Buchanan relates that the Scots mixed a quantity of the juice of belladonna with the bread and drink which,

---

*Another fanciful origin for the name, which signifies "beautiful woman," is, that it was bestowed in consequence of the use once made of its berries by

by their truce, they were to supply to the Danes with, which so intoxicated them, that the Scots killed the greater part of Sweno's army
while asleep.

The effects of belladonna on the brain are well described by Dr.
Winslow, than whom no better authority can be desired.

> One of the marvellous effects of continued doses is the production
> of a singular psychological phenomenon. A delirium supervenes,
> unaccompanied by any fantasia, or imaginary illusion, whose marked
> characteristic is somnambulism. An individual who has taken it in
> several doses seems to be perfectly alive to surrounding objects, his
> senses conveying faithfully to the brain the impressions that they
> receive; he goes through his usual avocations without exhibiting
> any unwonted feeling, yet is he quite unconscious of his existence,
> and performs mechanically all that he is accustomed to do, answers
> questions correctly, without knowing from whom or from whence
> they proceed, looks at objects vacantly, moves his lips as if convers-
> ing yet utters not a sound, there is no unusual state of the respira-
> tory organs, no alteration of the pulse, nothing that can bespeak
> excitement. When this state of somnambulism passes away, the in-
> dividual has not the slightest recollection of what has occurred to
> him; he reverts to that which immediately preceded the attack, nor
> can any allusion to his apparent reverie induce him to believe that
> he has excited any attention. The case of the tailor who remained
> on his shopboard for fifteen hours, performing all his usual avoca-
> tions, sewing with great apparent earnestness, using all the gestures
> which his business requires, moving his lips as if speaking, yet the
> whole time perfectly insensible, has been frequently quoted. It was
> produced by belladonna.

The use of this plant has been recommended as a preventive of
scarlatina. An instance is recorded of a family consisting of eleven
persons who took it for this purpose, in small quantities, twice a day.
Five of these persons were domestics. On the fourth day, almost all of
them became under the influence of the drug, two or three of them
very slightly, simply complaining of having the vision disturbed by
objects which they in vain attempted to remove, for they were fully
persuaded that they existed. Two had singular fits of laughter which

nothing could control. All complained of being in an unusual state. The servants were all of them able to go through their work, but all seemed to act mechanically, each independent of the other. Of this the most ludicrous example was in the course of the fourth evening. A carriage arrived at the street door, and the street bell was rung with considerable violence. They all immediately left their business, quietly walked up stairs as if they had not the slightest idea that they were all upon the same errand. They went to the door; two of them, however, only opened it; one of these walked away without waiting to know what was the reason of the ringing, and the other seemed not disposed to trouble himself with anything beyond the opening and shutting of the door. On the discontinuance of the medicine, they all soon returned to their usual state, and two of them had scarlatina, though only in a mild form.

From this descriptive account of the action of belladonna, and its singular effects upon the mind, approaching to a form of insanity, it will appear strange that this drug should be recommended by Hahnemann and his followers for the cure of insanity. But this is the very principle upon which that school operates.* That drug which produces, in its effects the worst forms of mania, is the best adapted for its cure. We are not, however, either apologists, exponents, or opponents of homœopathy, and will leave its supporters to champion their own cause.

Henbane (*Hyoscyamus niger*) is another of these powerful narcotic agents, educing symptoms analogous to insanity. In small doses, its effect is to produce a pleasant sleep and soothe pain. In larger doses, the effects are extremely deleterious. Two soldiers who ate the young shoots dressed with olive oil, became giddy and stupid, lost their speech, had a dull and haggard look. The limbs were cold and palsied, and a singular combination of delirium and coma manifested itself. As the palsy and somnolency decreased, the delirium became extravagant. Others who partook of the same species of plant by mistake were affected in a similar manner. Several were delirious and danced about the room like maniacs, and one appeared as if he got drunk. A French physician gives an account of nine persons who were nearly poisoned

---

*"Similia similibus curantur."

by eating the roots of henbane. The effects of this poison were horrible in the extreme; in five, out of the nine, it produced raving madness. The madness of all these was so complete, and their agitation so violent, that in order to give one of them an antidote, six strong men had to be employed to hold him down, while his teeth were being separated to pour down the remedy. For two or three days after their recovery, every object appeared to them as red as scarlet.

Henbane, which is often administered as a substitute for opium, and in the East occasionally mixed with it, has the extraordinary faculty of producing jealousy. Many authenticated cases are recorded of the power of the leaves, and the fumes of the seeds, over the more intense passions. A disposition to quarrel and fight is decidedly produced. One case is that of a young couple, who had married from affection, had lived upon terms of the most perfect mutual regard—indeed, had been noticed for the warmth and strength of their attachment; but suddenly, to the surprise of the surrounding neighbours, their harmony was not only interrupted, but they became bitter antagonists, fighting and beating each other most unmercifully. What seemed most surprising was, that in one particular room appeared to spring their most determined quarrels, and that they soon subsided elsewhere. This mystery was at length explained, and their days of happiness restored, by the discovery that to the effects of a considerable quantity of henbane, stored up for drying, their miseries were owing, and on the removal of this, the source of their feuds appeared to vanish. Hahnemann, as might be expected, considers this as one of the most potent medicines for the cure of jealousy, since it is so effective in causing it.

The leaves of the three plants lately noticed—namely, thorn-apple or stramonium, belladonna, and henbane—are made up in the form of cigarettes; and the first of these also as cigars, to be smoked by asthmatic persons, for their soothing and sedative effects. These are all made and consumed extensively on the continent, and may be procured in many parts of London. They have also been recommended to those *not* asthmatical, as pleasant, harmless, and containing all the narcotising influences of a good cigar. They may be considered as truly narcotic substitutes for tobacco; but at the present rate at which they are sold, although not subject to either customs or excise, there is but little fear of their interfering prejudicially with the sale of the genuine article. In face of the facts already detailed, a good amount of courage seems necessary to make the attempt, lest they should prove cumulative in their action. Dr. Christison says, when writing of these narcotics, "The action of such poisons is not always, however, entirely thrown away; they still produce some immediate effect; and further, by being frequently taken, they may slowly bring on certain diseases, or engender a predisposition to disease. A very singular exception to this rule prevails in the instance of tobacco, which, under the influence of habit, may be smoked daily to a considerable amount, and, so far as appears, without any cumulative effect on the constitution, like that of opium-eating or drinking spirits."

It does not appear that hitherto the leaves of the purple foxglove (*Digitalis purpurea*) has been used in the same form, or for any other than purely medicinal purposes; but it possesses narcotic powers equal to the others, and in excess, produces equally fatal results, such as delirium, convulsions, and insensibility. A fatal case which occurred in 1826 became the ground of a criminal trial, in which death took place in twenty-two hours, having been preceded by convulsions and insensibility.

An enumeration of the various other narcotics which enter into combination with other substances in the production of beverages, such as the hop and its substitutes, forming no part of the plan of this work, would be uninteresting without further details. Nor would a list of such narcotics as are used merely in materia medica answer any useful end. Fuller particulars would only convert this into a toxicological treatise, interesting to none but medical students, for whom ample information is provided in the libraries to which they have access.

CHAPTER XXV

THE EXILE OF SIBERIA

Vilibus ancipites fungi ponentur amicis;
Boletus domino.

Juvenal

The rage for scampering half over the world in search of the picturesque has scarcely got far enough to tempt any, except a stray traveller or two, into the chilly regions of Siberia and Kamtschatka, and in these exceptional cases, perhaps, more from force than choice. These are regions, therefore, concerning which our information is remarkably limited. It is true that Captain Cochrane informs us that he married a wife from Kamtschatka—a virtuous maiden, who knew more of that region, perhaps, than he or she cared to tell; for the one tells us very little, and the other nothing, of yon strange land, with an almost unpronounceable name. We are told, moreover, that the capital is called by the names of St. Peter and St. Paul. Fearing lest one patron saint should not be sufficient to immortalize the metropolis of all the Kamtschatkas, the founders and inhabitants have wisely adopted two. This city also is stated to contain forty-two dwellings, besides fifteen edifices belonging to the government, an old church, and the foundation of a new one. The winters are declared to be mild, compared with those of Siberia; but even these are not very inviting, as the snow lies on the ground seven or eight months, and the soil, at the depth of twenty-four to thirty inches, being frozen at all seasons. Potatoes never ripen, cabbages never come to a head, and peas only flower. But the gallant captain adds: "I am certainly the first Englishman that ever married a Kamtschatdale, and my wife is undoubtedly the first native of that peninsula that ever visited happy Britain."

In such a land, there is little hope of cultivating poppy, tobacco, betel, coca, hemp, or thorn-apple; and the poor native would have been compelled to have glided into his grave without a glimpse of Paradise beforehand, if, on the one hand, the kindly Russian pedlar had not found a way to smuggle a little bad spirits into the country, to the great annoyance of all quietly-disposed persons, or, on the other, nature had not promptly supplied an indigenous narcotic, in the form of an unpretending-looking fungus or toadstool, to stimulate the dormant energies of the dwellers in this region of ice and snow.

That some kinds of mushrooms are poisonous is a truth of which every farm labourer seems aware. But that some of those which have been reputed poisonous are inert, is beyond their philosophy, and only receives at present the sanction of some of the more scientific, who have directed their studies thitherward. The fly agaric is one of those justly-reputed poisonous species, occasionally found in this country, but which grows plentifully in Kamtschatka and Siberia. A recent author of an account of Russia states, "that mushrooms virulently poisonous in one country are eaten with safety in another, is well known in other cases, as, for instance, in that of the fly mushroom (*Amanita muscaria*), which is common in England, and always poisonous there, while in Kamtschatka it is used as a frequent article of food." Then he inquires into the reasons wherefore this should be the case:

> It is not enough to say that difference of soil and climate explain the mystery; for though we know that culture changes the properties of plants, converting what is poisonous in the wild state into a wholesome esculent when raised in the garden—as in the case of the common celery, for example—yet throughout the whole of the vegetable kingdom we find almost no other instance of a plant which is poisonous in one country becoming wholesome, without culture, when transplanted to another, and left entirely to itself, and in both placed in apparently the same circumstances as to soil, etc. After all, a great part of the secret may lie, not in the plant, but in the mode of preparing it for the table. So far as we can judge, the Russian cook, on first cutting up these spoils of the forest, makes a much

more copious use of salt than is done with us; and the efficacy of
this agent in deadening the poisonous quality, is sufficiently proved
by the melancholy case recorded in medical treatises, of a French
officer and his wife, both of whom died in thirty-two hours after
eating certain mushrooms, while the person who supplied them,
and his whole family, made a hearty and wholesome meal from the
same gathering.

In this case, it appears that while the former took them without addition, the latter first salted them strongly, and then squeezed them well
before using them. M. Roques says distinctly that this plant has not its
poisonous properties modified by any climate. The Czar Alexis lost
his life by eating this mushroom. The details of its effects upon the
Kamtschatkans by Krascheminikow, in his natural history of that country are explicit, respecting the delirious intoxication induced by it.
Gmelin and Pallas also equally certifying its intoxicating powers.
Roques reports seven different sets of observations respecting its deleterious effects on man.

Unless we accept some such explanation of the phenomena as
this, how can we reconcile the fact of their being eaten by the Russians without injury, whilst, on the authority of Dr. Christison, we
have such a fatal case as the following, from eating the same kind of
fungus, the growth of the same country and climate. Several French
soldiers in Russia ate a large quantity of *Amanita muscaria*, some were
not taken ill for six hours and upwards. Four of them who were very
powerful men thought themselves safe, because, while their companions were already suffering, they themselves felt perfectly well and
refused to take emetics. In the evening they began to complain of
anxiety, a sense of suffocation, frequent fainting, burning thirst, and
violent gripes. The pulse became small and irregular, and the body
bedewed with cold sweat, the lineaments of the countenance were
singularly changed, the nose and lips acquiring a violet tint, they
trembled much, the belly swelled, and a profuse diarrhœa followed.
The extremities soon became livid and cold, and the pain of the ab-
domen intense, delirium ensued, and all the four died. Two of the
others suffered coma for twenty-four hours.

This proves that the mushroom in question is possessed of undoubtedly poisonous properties, which are fatal in their effects, unless
counteracted or dispelled by the method of preparing them for the

table. That this method is known to the Russians and to some other nations, and is believed to consist in well saturating the fungi with salt before cooking them. The Muscovite seems to have no greater dread of ill effects from the fly agaric than has the Brazilian from his cassava or mandioca flour, which is prepared from the equally poison-ous root of the mandioca plant, the deleterious qualities of which are destroyed by the heat used in its preparation. Dr. Pouchet of Rouen seems to have clearly proved that the poisonous property of the fly agaric and *a venenata*, may be entirely removed by boiling them in water. A quart of water in which five plants had been boiled for fif-teen minutes, killed a dog in eight hours; and, again, another in a day; but the boiled fungi themselves had no effect at all on two other dogs; and a third which had been fed for two months on little else than boiled amanitas, not only sustained no harm, but actually got fat on the fare.* Pouchet is inclined to think that the whole poisonous plants of the family are similarly circumstanced.

The most singular circumstance connected with the history of this fungus, is the place it occupies as a substitute for those narcotics known in other parts of the world, and while an ungenial northern climate fails to produce. What the coca is to the Bolivian, and opium to the Chinese—the areca to the Malay, and haschisch to the African—the tobacco to the inhabitants of Europe and America, and the thorn-apple to those of the Andes—is the fly agaric to the natives of Siberia and Kamtschatka. Why it has been called by this name has arisen from its use as a fly poison. Never having seen those dipterous insects while under its influence, we cannot detail the symptoms it produces.

This poisonous fungus has some resemblance to the one generally eaten in this country, yet there are also striking points of difference. As, for instance, the gills are white instead of pinkish red, inclining to brown, and the cap or pileus, which is rather flat, is generally of a livid red colour, sprinkled with angular lighter coloured worts. These are distinctions broad enough to prevent any one having the use of his eyes, and who has ever seen the edible mushroom being deceived into belief that the fungus thus briefly described is identical with the deli-cacy of our English tables.

---

*Journ. de Chem Méd., 1839, p. 322.

These fungi are collected by those who indulge in them narcoti-
cally, during the hot, or rather summer months, and afterwards hung
up to dry in the open air. Or they may be left to ripen and dry in the
ground, and are afterwards collected. When left standing until they
are dried, they are said to possess more powerful narcotic properties
than when dried artificially. The juice of the whortleberry in which
this substance has been steeped, acquire thereby the intoxicating properties of strong wine.

The method of using this singular substance is to roll it up in the
form of a bolus and swallow it without any mastication, as one would
swallow a large pill. It is swallowed thus on principle, not that its
flavour would be unpleasant, as compound colocynth might be when
masticated, but because it is stated to agree ill with the stomach when
that operation is performed. Nature is jealous of her rights, and it would
appear from experience, that the gastronomic regions expect to receive all other supplies well triturated, except these—amanita and
pill colocynth—which are both expected equally alike to arrive at the
regions below without mutilation.

A day's intoxication may thus be procured at the expense of one
good sized bolus, compounded of one large or two small toadstools;
and this intoxication is affirmed to be, not only cheap, which is a
consideration, but also remarkably pleasant. It commences an hour or
so after the bolus has been swallowed.

The effects which this singular narcotic produces are, some of
them, similar to that produced by intoxicating liquors; others resemble
the effects of haschisch. At first, it generally produces cheerfulness,
afterwards giddiness and drunkenness, ending occasionally in the
entire loss of consciousness. The natural inclinations of the individual become stimulated. The dancer executes a *pas d'extravagance*,
the musical indulge in a song, the chatterer divulges all his secrets,
the oratorical delivers himself of a philippic, and the mimic indulges
in caricature. Erroneous impressions of size and distance are common occurrences, equally with the swallower of amanita and hemp.
The experiences of M. Moreau with haschisch are repeated with the
fungus-eaters of Siberia; a straw lying in the road becomes a formidable object, to overcome which, a leap is taken sufficient to clear a
barrel of ale, or the prostrate trunk of a British oak.

But this is not the only extraordinary circumstance connected therewith. There is the property imparted to the fluid excretions, of rendering it intoxicating, which property it retains for a considerable time. A man having been intoxicated on one day, and slept himself sober by the next, will, by drinking this liquor, to the extent of about a cupfull, become as intoxicated thereby as he was before. Confirmed drunkards in Siberia preserve their excretionary fluid as a precious liquor, to be used in case a scarcity of the fungus should occur. This intoxicating property may be again communicated to every person who partakes of the disgusting draught, and thus, also, with the third, and fourth, and even the fifth distillation. By this means, with a few boluses to commence with, a party may shut themselves in their room, and indulge in a week's debauch at a very economical rate. This species of "sucking the monkey" is one that Mungo never contemplated. Persons who are fond of getting liquor at the expense of others take every opportunity of "sucking the monkey," which process has been thus explained. It consists in boring a hole with a gimlet in a keg or barrel, and putting a straw therein, to suck out any quantity, at any given time. Persons who are accustomed to receive real Devonshire cider, or genuine Wiltshire ale, or the pure Geneva, in London experience the liberties those take who "suck the monkey," but either liberally diminishing the quantity, or diluting it with water on the road, so as to make the quantity what the quality should be. It is said that the origin of the term "sucking the monkey" is derived from the prolific invention of a black, who, in order to find an excuse to the captain for his being caught lying with a favourite monkey so often near the rum puncheons on board, from which he daily drank, said—"Massa, you ask what Mungo do here?—do here, massa? You say monkey hab de milk ob human kindness, massa. Mungo like dat milk, massa, and Mungo suck de monkey, massa. Dat's all."

Chemical investigations have not yet been directed into the channel leading towards the elucidation of the mysteries of these poisonous fungi, and hitherto we know of no experiments having been made with a view to ascertain whether any of our indigenous fungi, other than the one already referred to, can be used in the same way, and with the same results, as we have described. Doubtless such experiments would be successful, so far as realizing the results, since one of the effects produced by eating poisonous fungi is narcotic in its character. M. Letellier found in certain of these fungi a chemical principle which is fixed, and resists drying, and which he calls Amanitine. Its effects on animals appear to resemble considerably those of opium.* Dr. Christison states that "the symptoms produced by them in man are endless in variety, and fully substantiate the propriety of arranging them in the class of narcotico-acrid poisons. Sometimes they produce narcotic symptoms alone, sometimes only symptoms of irritation, but much more commonly, both together." A person gathered in Hyde Park a considerable number of mushrooms, which he mistook for the species commonly eaten, stewed them, and proceeded to eat them; but before ending his repast, and not more than ten minutes after he began it, he was suddenly attacked with dimness of vision, giddiness, debility, trembling, and loss of recollection. In a short time he recovered so far as to be able to go in search of assistance. But he had hardly walked 250 yards when his memory again failed him, and he lost his way. His countenance expressed anxiety, he reeled about, and could hardly articulate. He soon became so drowsy, that he could be kept awake only by constant dragging. Vomiting was produced; the drowsiness gradually went off, and next day he complained merely of languor and weakness.

The smoke of the common puff-ball when burnt, has been used to stupify bees when their hive was about to be robbed; and similar narcotic effects have been observed in other animals when subjected to its fumes. The action bears a resemblance to that of chloroform by producing insensibility to pain. If future generations do not deem it desirable to indulge in a narcotic of this kind for the purpose of producing pleasurable sensations, or to smother the carking cares of life, yet they may learn more than we at present know of the peculiar characteristics which distinguish this from all the others of the "Seven Sisters of Sleep."

Night draws on apace; let us gather together all the straggling members of the family, sweep up the crumbs, call in the cat, bar the door, wind up the clock, and go to bed—

To sleep, perchance to dream.

---

*Archives Gén. de Méd., t.xi., p. 94.

CHAPTER XXVI

⍥DDS AND ENDS

And our poor dream of happiness
Vanisheth, so
Farewell.

Motherwell

After a feast, the prudent and thrifty housewife will gather up the fragments that remain, if for no other purpose than to distribute them amongst the poor.

It was the constant habit if a certain elderly man of business, so long as he could stoop for the purpose, to pick up and stow away every pin and scrap of paper, or end of string, which he saw lying about on his premises. And when he could bend no longer to perform the operation himself, he would stand by the truant fragment, and vociferate loudly for one of his apprentices to come and "gather up the cord and string," adding " 'tis a pity they should spile."

Approaching to the conclusion of our task, we have followed the old gentleman's advice, and collected the odd pieces that have fallen to the ground in the course of our work, convinced that thrift is praiseworthy, and although only "Odds and Ends," there may be enough of interest in them to warrant you in adding " 'tis a pity they should spile."

Tobacco ends in smoke. We began with the former, it is but a natural consequence that we should end with the latter. Somewhere we have read a "smoke vision of life." Some people have but a smoky or foggy vision of life—they live in a mist, and die without being missed. Forgive the transgression, good friend, the obscurity of the subject is to blame, and the pun was written before we had made ourselves aware of its presence. Let it pass on, it will soon be lost in the smoke. An old piper believes that there is generally something racy, decided, and original in the man who both smokes and snuffs. Outwardly, he may have a kippered appearance, and his voice may grate on the ear like a scrannel pipe of straw, but think of the strong or beautiful soul that body enshrines! Do you imagine, oh, lean-hearted member of the Anti-Snuff and Tobacco Club, that the dark apostle standing before us will preach with less power, less unction, less persuasive eloquence, because he snuffs over the psalm book, and smokes in the vestry between the forenoon and afternoon service? Does his piety ooze through his pipe, or his earnestness end in smoke? Was Robert Hall less eloquent than Massillon or Chalmers, because he could scarcely refrain from lighting his hookah in the pulpit? Answer us at your leisure—could Tennyson have brought down so magnificently the Arabian heaven upon his nights; dreamed so divinely of Cleopatra, Iphigenia, and Rosamond; pictured so richly the charmed sleep of the Eastern princess in her enchanted palace, with her "full black ringlets downward rolled"; or painted so soothingly the languid picture of the Lotos-eaters, if he had never experienced the mystic inspiration of tobacco? Could John Wilson—peace to his princely shade—have filled his inimitable papers with so much fine sentiment, radiant imagery, pathos, piquancy, and point, without the aid of his silver snuff-box? Deprive the Grants and Macgregors of their mulls and nose-spoons of bone, and you cut the sinews of their strength—you destroy the flower of the British army. Pluck the calumet of peace from the lips of the red Indian, and in the twinkling of an eye your beautiful scalp will be dangling at his girdle. Tear his "gem adorned chibouque" from the mouth of the Turk, and the Great Bear by to-morrow's dawn will be grinning on his haunches in Constantinople. Clear Germany of tobacco smoke, and Goethe would groan in his grave, Richter would revisit the glimpses of the moon, philology would fall down in a fatal fit of apoplexy over the folios of her fame, and poetry would shriek her death-shriek to see the transcendental philosophy expire. Shake the quids from the mouths of the merry mariners of England—cast their pig-tail upon the waters, and commerce would become stagnant in all our ports—our gallant war-fleet would rot at its stations, and Britain would never boast the glories of another Trafalgar. Tell Yankeedom that smoking is no more to be permitted all over the world, under penalty of death, and soon the melancholy pine forests would wave over the dust of an extinguished race. In fine, were the club to which you belong to succeed in its attempt, which it can-not, the earth would stand still, like the sun of old upon Gibeon, and the moon in the valley of Ajalon, and the planets would clothe them-selves with sackcloth for the sudden death of their sister sphere!

There is extant, in an old work written three centuries since, a curious paragraph, which we had well nigh forgotten. It refers to Canada. "There groweth a certain kind of herbe, whereof in summer they make great provision for all the yeere, and only the men use of it; and first they cause it to be dried in the sunne, then wear it about their neckes, wrapped in a little beaste's skinne, made like a little bagge, with a hol-low peece of stone or wood like a pipe; then, when they please, they make a poudre of it, and then put it in one of the ends of the said cornet or pipe, and laying a coal of fire upon it, at the other end suck so long, that they fill their bodies full of smoke, till that it cometh out of their mouth and nostrils, even as out of the tonnell of a chimney."

Methinks it had been well had every Canadian been also favoured with a Saint Betsy, as a companion in life, otherwise there had been fire as well as smoke. It is now some time since the inimitable *Punch* introduced Saint Betsy to the world, and that she may not altogether be excluded from our future "fireside saints," we will give her legend a place in our "Odds and Ends."

> St. Betsy was wedded to a knight who sailed with Raleigh, and had brought home tobacco, and the knight smoked. But he thought that St. Betsy, like other fine ladies of the court, would fain that he should smoke out of doors, nor taint with tobacco smoke the tapestry, where-upon the knight would seek his garden, his orchard, and, in any weather, smoke *sub Jove*. Now it chanced, as the knight smoked, St. Betsy came to him and said, "My lord, pray ye come into the house"; and the knight went with St. Betsy, who took him into a newly cedared room, and said, "I pray my lord henceforth smoke here, for is it not a shame that you, who are the foundation and prop of your house, should have no place to put your head into and smoke?" And St. Betsy led him to a chair, and with her own fingers filled him a pipe; and from that time the knight sat in the cedar chamber and smoked his weed.

No pipe, no smoke, no dreams! Never again, on a beautiful

summer's day would two young Ottoman swains sit smoking under a tree, by the side of a purling stream, hearing the birds sing, and seeing the flowers in bloom, to become the actors in a scene like that described in one of their own songs. By and bye came a young damsel, her eyes like two stars in the nights of the Ramazan. One of the swains takes his pipe from his mouth, and "sighing smoke," gazes at her with delight. The other demands why his wrapt soul is sitting in his eyes, and he avows himself the adorer of the veiled fair. "Her eyes," says he, "are black, but they shine like the polished steel, nor is the wound they inflict less fatal to the heart." The other swain ridicules his passion, and bids him re-fill his pipe. "Ah, no!" cries the lover, "I enjoy it no more; my heart is as a fig thrown into a thick leafy tree, and a bird with bright eyes has caught it and holds it fast."

Hearken to the story of Abou Gallioun, the father of the pipe-bowl, and then laugh if you will at the votaries of the marvellous weed. A mountaineer of Lebanon, a man young and tall, and apparently well to do, for his oriental costume was rich and elegant, established himself at Tripoli, in Syria. He resided at an hotel, and astonished every one with a bowl at the end of his pipe stem of enormous dimensions. Some days after his arrival he was seen to seat himself at the corner of a street, to rest the bowl of his pipe on the ground, and to take from his pocket a little tripod and coffee-pot. Having filled his coffee-pot, he put the tripod upon the bowl of his pipe, and stood his coffee-pot thereon. He then proceeded to smoke, and at the same time to boil the water for his coffee. This sight caused the passers-by to stop, and a crowd collected in the street so as to obstruct the throughfare. The police came to clear the passage, and, at the same time, the Pacha was informed of the circumstance, and consulted as to what should be done. The Pacha gave instructions that the stranger did harm to no one, he was to be allowed to make his coffee in the street, for the street was open to all, hoping that when it rained he would certainly go away. The police were, therefore, ordered to prevent any crowding around the mountaineer, and to take especial care that he received no insult, lest he should then complain to the Emir of the mountain of his ill-treatment. The mountaineer having heard of the instructions of the Pacha, continued to drink his coffee and smoke his pipe as before, in the presence of numbers of curious spectators. This exhibition continued daily, till the news penetrated into the harems, and the women came to see a man make his coffee upon the bowl of his pipe—a thing they had never before heard of, and which, till now, had never occurred.

The mountaineer loved to converse with the passers-by , when he told them that his pipe served him also at home for his baking oven, and that he had no other chafing dish in winter; that he filled the bowl twice a day, in the morning on rising, and in the evening on going to rest, to last him through the night; that he stopped very little, and during the night drank five or six cups of coffee. This stranger was surnamed Abou Gallioun, "father of the pipe-bowl," and is still known by that name in Tripoli when they speak of him and his extravagance.

In general, the pipe bowls are of a certain size, so that they may last at least a quarter of an hour, and with slow smoking they will last half an hour. The tobacco does not burn rapidly if the smoker does not pull hard—this quiet kind of smoking generally characterizes the grave orientals. Their pipes are seldom extinguished of themselves unless laid down, because the tobaccos of the East have more body than other tobaccos. Abou Gallioun might then always rest assured that his pipe would never go out, although he held long conversations by day, and rose occasionally at night to take his coffee.

Tobacco is stated to have been imported into the Celestial empire by the Mantchoos; and the Chinese were much astonished when they first saw their conquerors inhaling fire through long tubes and "eating smoke." By a curious coincidence this plant is called by the Mantchoos *tambakou*; but the Chinese designate it simply by the word meaning "smoke." Thus they say they cultivate in their fields the "smoke-leaf," they "chew smoke," and they name their pipe the "smoke-funnel."

The old proverb that "smoke doth follow the fairest," is thus commented upon: "Whereof Sir Thomas Brown says, although there seems no natural ground, yet it is the continuation of a very ancient opinion, as Petrus Victorious and Casaubon have observed from a passage in Athenæus, wherein a Parasite thus describes himself—

To every table first I come,
Whence Porridge I am called by some;
Like whips and thongs to all I ply,
Like smoak unto the fair I fly.

There is extant in the East, an Arabian tale concerning the Broken Pipe of Saladin, which is taken from an author named Ali-el-Fakir, who lived in the times of Saladin, a tale which is often repeated among smokers in Syria. The Sultan, Salah-el-Din (called by us Saladin), was a great warrior, a lover of the harem, and at the same time pleasant. His court abounded with officers, servants, and slaves. Among his servants, who could best amuse him in his leisure moments, was a simple man to whom he had confided the care of his pipes, and whom he had made his pipe-bearer. All the Sultan's pipes were of great value, owing to the oriental luxury which prevails in everything, and especially in everything belonging to the Sultan, who is considered the master of the world.

Saladin, in consequence of the climate of the south of Syria, generally passed his time in the gardens of Damascus, luxuriously seated upon rich Persian carpets and soft cushions, under a tree surrounded by his guards, and a numerous band of servants, who promptly obeyed his commands.

Under another tree, not far off, was the coffee-maker, ready to serve his master on the instant, for, like all other orientals, he was fond of this beverage; and Ramadan, the pipe-bearer, was commanded to be at hand, that he might execute his sovereign's orders.

Between the tree under which the Sultan was reposing, and that under which was the stove of the coffee-maker, stood another tree, to which was tied a watch-dog, who was only let loose at night.

Saladin said to Ramadan—"Take my pipe, fill it, and bring it to me directly." At that time tobacco was not smoked in the East, instead thereof they used Tè bégh. Ramadan hastened to obey his master, but the dog, not well knowing him, set to barking at him as he passed on his way to the coffee-maker's stove for the purpose of preparing there the Sultan's pipe, and in return Ramadan shook his fist at him. When the pipe-bearer came back, the dog, recognizing in him the man who had lately menaced him, not being securely tied, loosened himself and sprang at him. Ramadan used the pipe to defend himself, the dog was beaten back, but the bowl, the stem, and the rich mouth-piece of the pipe were all broken in the encounter.

The facts were related to Saladin, who immediately ordered the dog to be summoned before him. The animal said nothing while Ramadan was continually charging him with the blame. "Thou seest," said the Sultan, "that the dog appears docile. If thou hadst not threatened or frightened him he would have said nothing to thee. Thou shalt be tied up as the dog was, and the dog shall dwell with me."

The guards chained up poor Ramadan to the tree where the dog had been fastened, and his appearance was very disconsolate. The dog became the favourite of the Prince, whom he recognized by his natural instinct, and for ever afterwards the Sultan swore by his dog.

The Mussulman delights in comparing the wisdom of this decision with the judgment of Solomon.

The recent remarks of one high in clerical authority, which came to light but too lately to have a more honourable position assigned them, must accordingly be scattered among the fragments. "Heaven forbid," writes the reverend gentleman,

> that I should ever see in England what I have more than once seen in France—a fine and gorgeously arrayed lady, with lavander coloured kid gloves, and a delicate little cigarette between her lips, expectorating in the most refined manner into a polished spittoon, and accompanying her male friends in inhaling the fumes of this noxious weed! No, our ladies have not countenanced the custom by example, but they have fostered it, cherished it, promoted it by their too much good nature, and allowed their husbands, brothers, and sons, and perhaps, their intended husbands, to enjoy their cigars in their presence, and even in their houses.
>
> Oh horrible, most horrible!

Hearken still further.

> I don't scruple to confess that I sat down to the consideration of this subject strongly prejudiced, personally and socially, against this evil practice; but I rise from the examination of the facts of the case surprised at the magnitude of the abomination to which it gives rise. I cordially throw any influence I possess into the scale of those who are labouring to promote the total abolition of the custom among us, and I earnestly entreat all who think with me to exert their utmost efforts to stay the plague.

King James is dead, poor man, otherwise this worthy dean, most assuredly, would soon have become a bishop. How unfortunate a circumstance it is that wise men will be born at a time when the generation who would have appreciated them most, is either extinct or in embryo.

We remember to have once heard an equally estimable clerical gentleman declare that he thought those words of Longfellow's very descriptive of the effects of his customary "whiff":

And the night shall be filled with music,
And the cares that infest the day,
Shall fold their tents like the Arabs,
And silently steal away.

With a fable of Krummacher's, let this basket of fragments be filled, and finished.

"The angel of sleep and the angel of death, fraternally embracing each other, wandered over the earth. It was eventide. They laid themselves down beside a hill not far from the habitations of men. A melancholy silence reigned around, and the evening bell of the distant hamlet had ceased.

Silently and quietly, as is their wont, the two kindly genii of the human race lay in confidential embrace, and the night began to steal on.

"Then the angel of sleep rose from his mossy couch, and threw around, with careful hand, the unseen grains of slumber. The evening wind bare them to the quiet dwellings of the wearied husbandmen. Now the feet of sleep embraced the inhabitants of the rural cots, from the hoary headed old man who supported himself on his staff, to the infants in the cradle. The sick forgot their pains, the mourners their griefs, and poverty its cares. All eyes were closed.

And now, after his task was done, the beautiful angel of sleep lay down again by the side of his sterner brother. When the morning dawn arose, he exclaimed in joyous innocency—"Men praise me as their friend and benefactor. Oh what a bliss it is, unseen and secretly to befriend them! How happy are we, the invisible messengers of the good God! How lovely is our quiet vocation!"

Thus spake the friendly angel of sleep. And the angel of death sighed in silent grief; and a tear, such as the immortals shed, trembled in his great dark eye. "Alas!" said he, "that I cannot as thou, delight myself with cheerful thanks. Men call me their enemy and pleasure spoiler."

"Oh, my brother," rejoined the angel of sleep, "will not the good also, when awakening, recognize in thee a friend and benefactor, and thankfully bless thee? Are not we brothers and messengers of one Father."

Thus spake he, and the eyes of the angel of death sparkled, and more tenderly did the brotherly genii embrace each other.

APPENDIX

Table 1. Chronology of Tobacco

A.D.	
1496	Romanus Paine published the first account of tobacco, under the name *cohoba*.
1519	Tobacco discovered by the Spaniards near Tabasco.
1535	Negroes cultivated it on the plantations of their masters. It was used at this time in Canada.
1559	Tobacco introduced into Europe by Hernadez de Toledo.
1565	Conrad Gesner became acquainted with tobacco. Sir John Hawkins brought tobacco from Florida.
1570	Tobacco smoked in Holland out of tubes of palm-leaves.
1574	Tobacco cultivated in Tuscany.
1575	First figure of plant in Andre Thevot's *Cosmographie*.
1585	Clay pipes noticed by the English in Virginia. First clay pipes made in Europe.
1590	Schah Abbas, of Persia, prohibited the use of tobacco in his empire.
1601	Tobacco introduced into Java. Smoking commenced in Egypt about this time.
1604	James I laid heavy imposts on tobacco.
1610	Tobacco-smoking known at Constantinople.
1615	Tobacco first grown about Amersfort, in Holland.
1616	The colonists cultivated tobacco in Virginia.
1619	James I wrote his *Counterblast*.

(Table 1—continued)

1619	Sale of tobacco prohibited in England till the custom should be paid, and the royal seal affixed.
1620	Ninety young women sent from England to America, and sold to the planters for tobacco at 120 lbs. each.
1622	Annual import of tobacco into England from America, 142,085 lbs.
1624	The Pope excommunicated all who should take snuff in church. King James restricted the culture of tobacco to Virginia and the Somer Isles.
1631	Tobacco-smoking introduced into Misnia.
1634	A tribunal formed at Moscow to punish smoking.
1639	The Assembly of Virginia ordered that all tobacco planted in that and the succeeding two years should be destroyed.
1653	Smoking commenced at Apenzell (canton) in Switzerland.
1661	The police regulations of Berne made, and divided according to the Ten Commandments, in which tobacco was prohibited.
1669	Adultery and fornication punished in Virginia by a fine of 500 to 1000 lbs. of tobacco.
1670	Smoking tobacco punished in the canton of Glarus by fines.
1676	Customs on tobacco from Virginia collected in England, £120,000. Two Jews attempt the cultivation of tobacco in Brandenburg.
1689	Dr. J. F. Vicarius invented tubes containing pieces of sponge for smoking tobacco.
1691	Pope Innocent XII excommunicated all who used tobacco in St. Peter's Church at Rome.
1697	Large quantities of tobacco produced in the palantinate of Hesse.
1709	Exports of tobacco from America, 28,858,666 lbs.
1719	Senate of Strasburg prohibited the culture of tobacco.
1724	Pope Benedict XIV revoked Pope Innocent's Bull of excommunication.
1732	Tobacco made a legal tender in Maryland, at one penny per lb.
1747	Annual exports of tobacco to England from the American colonies, 40,000,000 lbs.

(Table 1—continued)

1753	The King of Portugal farmed out the tobacco trade for about £500,000. The revenue of the King of Spain from tobacco, £1,250,000.
1759	Duties on tobacco in Denmark amounted to £8,000.
1770	Empress of Austria derived an income of £160,000 from tobacco.
1773	Duties on tobacco in the two Sicilies, £80,000.
1775	Annual export of tobacco from the United States 1,000,000 lbs.
1780	King of France derived an income of £1,500,000 from tobacco.
1782	Annual export of tobacco during the Seven Years Revolutionary War, 12,278,504 lbs.
1787	Tobacco imported into Ireland, 1,877,579 lbs.
1789	Exports of tobacco from the United States, 90,000,000 lbs. Tobacco first put under the excise in England.
1820	Quantity of tobacco grown in France, 32,887,500 lbs.
1828	Tobacco revenue in the State of Maryland, £5,400.
1830	Revenue from tobacco and snuff in Great Britain was $2\frac{1}{4}$ millions of pounds
1834	Value of tobacco used in the United States estimated at £3,000,000.
1838	Annual consumption of tobacco in the United States estimated at 100,000,000 lbs.
1840	It was ascertained that 1,500,000 persons were engaged in the cultivation and manufacture of tobacco in the United States.

Table 2. Consumption of Tobacco

Countries	Average consumption of male population per head, over 18 years of age (pounds)	Net revenue from tobacco
Austria	6.75	£1,212,530
Zollverein	9.75	296,560
Steurverein, including Hanover and Oldenburg	12.50	12,420
France	5.40	3,058,356
Russia	2.50	284,280
Portugal	3.50	304,140
Spain	4.75	1,268,082
Sardinia	2.75	246,192
Tuscany	2.50	84,860
Papal States	2.00	297,252
Two Sicilies	—	168,422
Britain	4.10	5,272,471
Holland	8.25	6,210
Belgium	9.00	28,014
Denmark	8.00	10,488
Sweden	4.37	14,766
Norway	6.40	23,322
United States	7.60	—

Table 3. Duties on Importation of Tobacco

United States	30.0	per cent. ad valorem
Belgium	13.9	do.
Great Britain	933.3	do.
Hanover	9.6	do.
Holstein	10.0	do.
Holland	3.5	do.
Russia	161.0	do.
Switzerland	3.0	do.
Zollverein	45.0	do.

Table 4. Nett Profits of the French Regie on Tobacco, after Paying All
Expenses of Purchase, Transportation, Manufacture, and Sale (Showing the
Increased Consumption, in Decennial Periods, from 1811 to 1851)

	Years	Francs
	1811	26,000,000
	1821	42,219,604
	1831	45,920,930
	1841	71,989,095
	1851	92,233,729
Total gross revenue in 1857		185,000,000

Table 5. Consumption of Tobacco in Britain, with Rate of Duty and
Revenue Therefrom

Years	Consumption (pounds)	Duty (per pound)	Revenue	Population
1821	15,598,152	4s.	£3,122,583	21,282,903
1831	19,533,841	3s.	2,964,592	24,410,459
1841	22,309,360	3s.	3,580,163	27,019,672
1851	28,062,978	3s.	4,485,768	27,452,262
1856	32,579,166	3s.	5,216,770	*
1857	32,677,059	3s.	5,231,455	*
1858	34,110,850	3s.	5,272,471	*

* Owing to extensive emigration, especially from Ireland, the population must
be considered as but little above that of 1851.

Table 6. Consumption of Tobacco in the Austrian Empire

Years	Quantity consumed (pounds)
1850	34,457,513
1851	54,217,578
1852	61,805,697
1853	57,926,925
1854	62,020,333
1856	85,161,030

Table 7. Statement Exhibiting the Quantities of Tobacco Exported from the United States into the Countries Named, During 1855

Countries	Quantities (pounds)
Bremen	38,058,000
Great Britain	24,203,000
France	40,866,000
Holland	17,124,000
Spain	7,524,000
Belgium	4,010,000
Sardinia	3,314,000
Austria	2,945,000
Sweden and Norway	1,713,000
Portugal	336,000

Table 8. Disposition of Tobacco Growth of the United States in 1840 and in 1850, with the Home Consumption at Each Period

Years	Growth (pounds)	Exports (pounds)	Consumption (pounds)	Rate per head (ounces)
1840	219,163,319	184,965,797	34,543,557	$32^1/_2$
1850	199,532,494	122,408,780	81,933,571	56

Table 9. Statement Showing the Exports of Tobacco from America (United States) in Decennial Periods, from 1820 to 1850, and in 1855

Years	Quantity exported (hogsheads)
1820	66,000
1830	83,810
1840	119,484
1850	145,729
1855	150,213

Table 10. Analysis of Tobacco by Posselt and Reinmann

Nicotina	0.06
Concrete vegetable oil	0.01
Bitter extractive	2.87
Gum, with malate of lime	1.74
Chlorophylle	0.267
Albumen and gluten	1.308
Malic acid	0.51
Lignin and a trace of starch	4.969
Salts (sulphate, nitrate, and malate of potash, chloride of potassium, phosphate and malate of lime, and malate of ammonia)	0.734
Silica	0.088
Water	88.280
Fresh leaves of tobacco	100.836

Table 11. Return Showing the Quantity of Chests of Opium Exported by the East Indian Company between 1846 and 1858

Years	No. of chests
1846–47	22,468
1847–48	22,879
1848–49	33,073
1849–50	35,919
1850–51	32,033
1851–52	31,259
1852–53	35,521
1853–54	42,403
1854–55	49,979
1855–56	49,399
1856–57	66,305
1857–58	68,004

*Each chest of opium contains about 140 lbs.

Table 12. Amount of Income Derived by the East India Company from the Opium Monopoly

Years	Amount (£)
1840–41	874,277
1841–42	1,018,765
1842–43	1,577,581
1843–44	2,024,826
1844–45	2,181,288
1845–46	2,803,350
1846–47	2,886,201
1847–48	1,698,252
1848–49	2,845,762
1849–50	23,309,637
1850–51	3,043,135
1851–52	3,139,247
1852–53	3,717,932
1853–54	3,359,019
1854–55	3,333,601
1855–56	3,961,975
1856–57	3,860,390
1857–58	5,918,375

Table 13. Opium Statistics of Great Britain

Years	Imports (pounds)	Consumption (pounds)
1826	79,829	28,329
1827	113,140	17,322
1830	209,076	22,668
1833	106,846	35,407
1836	130,794	38,943
1839	196,247	41,632
1842	72,373	47,432
1845	259,644	38,229
1848	200,019	61,055

(Table 13—continued)

Years	Imports (pounds)	Consumption (pounds)
1849	105,724	44,177
1850	126,318	42,324
1851	118,024	50,682
1852	205,780	62,521
1853	159,312	67,038
1854	97,427	61,432
1855	50,143	34,473
1856	51,479	38,609
1857	136,423	56,174
1858	82,085	77,639

Table 14. Analysis of Opium, by Mulder

Morphia	10.842	4.106
Narcotina	6.808	8.150
Codeia	0.678	0.834
Narceine	6.662	7.506
Meconine	0.804	0.846
Meconic acid	5.124	3.968
Fat	2.166	1.350
Caoutchouc	6.012	5.026
Resin	3.582	2.028
Gummy extractive	25.200	31.470
Gum	1.042	2.896
Mucus	19.086	17.098
Water	9.846	12.226
Loss	2.148	2.496
Total	100.000	100.000

Table 15. Prisoners Sentenced by the Police to the House of Correction at Singapore

Class (Chinaman)[a]	Quantity of opium consumed daily (grains)	Number of years habituated	Trade	Monthly wages		Value of opium smoked monthly			Appearances
				s.	d.	£.	s.	d.	
1	60	10	Cooly	16	0	1	4	0	Heavy, listless, but not sleepy
2		b	—	—	—	—	—	—	Looks well and fat
3		b	—	—	—	—	—	—	Looks well, but not stout
4		b	—	—	—	—	—	—	Looks well
5	180	10	Planter	—	—	3	12	0	Looks well; given up smoking; drinks Tinco in arrack
6	90	12	—	—	—	1	10	0	Sickly, with cough
7	60	20	Cooly	16	0	1	4	0	Sickly, thin, and miserable looking
8	180	7	Planter	12	10	3	12	0	Sick and herpetic
9	90	6	—	20	0	1	10	0	Sickly looking, and complains
10	60	20	Cooly	16	0	1	4	0	Thin, sickly; complains of pain in the stomach
11	48	4	Cooly	16	0	0	16	4	Yellow, sickly; pain in the abdomen
12	300–350	16	Planter	—	—	£6 to £7			Thin, sickly; complains of cough
13	30	10	Cooly	16	0	0	12	0	Complains of pain in abdomen
14	90	6	—	16	0	1	10	0	Thin, but not sickly
15	60	16	Cooly	16	0	1	4	0	Thin, cough, and sickly
16		b	—	—	—	—	—	—	

(Table 15—continued)

Class (Chinaman)[a]	Quantity of opium consumed daily (grains)	Number of years habituated	Trade	Monthly wages		Value of opium smoked monthly			Appearances
				s.	d.	£.	s.	d.	
17	24	9 [b]	Cooly	16	0	0	10	0	Complains of pain in abdomen; does not look sickly
18	60–180	30	—	20	0	24s. to £3	12	0	Sickly looking; does not complain
19	36	5	—	24s. to 30s.		0	12	0	Diarrhœa, and complains
20	30	5	—	16	0	0	8	0	Complains, but does not look sickly
21	60	12	—	16	0	1	4	0	Complains, but does not look sickly
22	48	5	Cooly	12	0	1	0	0	Looks sickly, and complains
23		b	—	—		—	—	—	Looks sickly
24		b	—	—		—	—	—	Looks well
25		b	—	—		—	—	—	Looks well
26	60	15	—	16	0	1	4	0	Complains much, being without chandu
27		b	—	—		—	—	—	Looks well
28	36	6	—	12	0	0	15	0	Pale, sickly looking; complains much
29	48	5	Shop-keeper	—		1	0	0	Thin and sickly

[a]Besides which, there were 15 men in the hospital, of whom all smoked but one.
[b]Does not smoke.

Table 16. Opium Consumed by Fifteen Persons
from the Pauper Hospital, Singapore

	Quantity of opium consumed daily (grains)	Years habitu- ated	Monthly wages		Excess of expenditure over income	
			s.	d.	s.	d.
1	36	7	11	6	5	8
2	36	3	8	0	6	6
3	24	5	8	0	1	8
4	36	8	12	0	2	6
5	42	20	16	0	0	10
6	30	10	10	0	2	1
7	24	7	8	0	1	8
8	30	10	12	0	Income and expenditure equal	
9	24	5	8	0	1	8
10	30	10	8	0	4	0
11	30	8	12	0	Income and expenditure equal	
12	36	10	12	0	2	6
13	30	15	12	0	Income and expenditure equal	
14	30	25	12	0	"	
15	42	22	12	0	4	10

Table 17. Reports of Opium-Smoking in China

In the Chung-wan (centre bazaar) there are about 5,800 inhabitants.

The number that smoke opium merely because they like it are upwards of 2,600.

The number that smoke opium are upwards of 300.

In the Hah–wan (Canton bazaar) there are upwards of 1,200 inhabitants.

The number that smoke opium merely because they like it are upwards of 600.

The number that smoke opium are upwards of 100.

The number that died for cause of smoking opium very few.

(Signed) CHUNG-WAN & HAH-WAN TEAPOA'S REPORT.

Dated Yuet-man year, 11th month, 20th day

(December 29th, 1855).

The number of male residents at Sheong-wan are estimated as following: This year have ascertained the number of male residents are 13,000.

There are 3,000 opium-smokers; 300 smoke 8 mace a-day; 700 smoke 5 mace each day; 1,000 smoke 3 mace each day; the rest smoke 1 mace, more or less.

The number that smoke opium merely because they like it are upwards of 4,000.

The number that got sick for cause of opium-smoking went home, and did not die here.

(Signed) TEAPOA OF SHEONG-WAN TONG CHEW'S REPORT.

Dated December 29th, 1855.

By order, have ascertained the number of inhabitants of Tai-ping-Shan. There are upwards of 5,300 men.

The number that smoke opium because they like it are upwards of 1,200.

The number that smoke opium are upwards of 600.

The number that died for cause of opium-smoking very few.

(Signed) TAI-PING-SHAN TEAPOA'S REPORT.

Dated Yuet-man year, 11th month, 20th day

(December 29th, 1855).

By order, have ascertained that in Wan-tsai there are upwards of 1,600 inhabitants.

Those that smoke opium merely because they like it are upwards of 500 men.

Those that smoke opium are upwards of 200 men.

Those that died for cause of smoking opium, none.

Dated Yuet-man year, 11th month, 20th day

(December 29th, 1855).

(Table 17—continued)

By order, have ascertained that in Wang-nai-choon there are upwards of 200 men.

The number that smoke opium are upwards of 10 men.

The number that smoke opium merely because they like it are few only.

The number that died for cause of smoking opium, very few.

(Signed) WANG-NAI-CHOON TEAPOA'S REPORT.

Dated Yuet-man year, 11th month, 20th day

(December 29th, 1855).

By order, have ascertained the number of inhabitants of Ting-loong-chow (east point).

There are upwards of 2,500 inhabitants.

The number that smoke opium merely because they like it are upwards of 300.

The number that smoke opium are upwards of 100.

(Signed) TING-LOON-CHOW TEAPOA'S REPORT.

Dated Yuet-man year, 11th month, 20th day

(December 29th, 1855).

Table 18. Professor Johnston's Estimate of the Number of Persons Indulging in the Seven Principal Narcotics of the World

Tobacco	800,000,000
Opium	400,000,000
Hemp	200,000,000–300,000,000
Betel	100,000,000
Coca	10,000,000
Thorn–apple (no estimate)	Less than coca
Amanita	Less than coca

Table 19. Synopsis of Narcotics, with Their Substitutes

Vulgar name	Botanical name	Where used or cultivated	How used
		I. Tobacco	
Virginian tobacco	Nicotiana tabacum	U. States	Smoked and chewed
Orinoco "	" macrophylla	"	"
European "	" rustica	Europe	Smoked
Javanese "	" var.	Java	"
Billah "	" var Asiatica	Malwa	"
Guzerat "	" var	Guzerat	"
Chinese "	" var Chinensis	China	"
Thibetian "	" var	Thibet	"
Persian "	" Persica	Persia	"
Latakia "	" var.	Syria	"
Djiddar "	" crispa	"	"
Indian "	" quadrivalvis	N. America	"
" "	" multivalvis.	"	"
" "	" nana.	Rocky Mts.	"
Cuban "	" repanda.	Cuba	"
Columbian "	" loxensis.	America	"
Brazilian "	" glauca.	Brazil	"
Peruvian "	" andicola.	Andes	"
Coltsfoot leaves	Tussilago farfar.	Europe	Smoked for tobacco
Yarrow "	Achillœa millefolium	"	"
Rhubarb "	Rheum emodi, etc.	Himalayas	"

(*Table 19—continued*)

Vulgar name	Botanical name	Where used or cultivated	How used
Bogbean	Menyanthes trifoliata	Britain	"
Sage "	Salvia officinalis	Europe	"
Mountain tobacco	Arnica montana	Switzerland	"
Black holly	Ilex vomitoria	N. America	"
Stag's horn sumach	Rhus typhina	Mississippi	"
Copal sumach	Rhus copallina	"	"
Water-lily leaves	Nelumbium speciosum	China	Mixed with tobacco
Pucha-pat	Marrubium odoratissimum	India	Mixed with tobacco
Tombeki	Lobelia sp.	E. Asia	Smoked as tobacco
Indian tobacco	Lobelia inflata	N. America	
Maize husks	Zea Mays	U. States	Patented for cigars
Birch bark	Betula excelsa	N. Brunswick	Mixed with tobacco
Willow leaves	Salix sp.	N. America	Smoked as tobacco
Bearberry leaves	Arctostaphylus uva-ursi	Chenook Ind.	Mixed with tobacco
Pimento berries	Eugenia pimento	W. Indies	Smoked
Cascarilla bark	Croton eleuteria	"	Mixed with tobacco
Polygonum leaves	Polygonum hispida	S. America.	Smoked
Camphor leaves	Tarchonanthus camphoratus	Cape	"
Wild dagga	Leonotis leonurus	"	"
	Leonotis ovata	"	"
Culen	Psoralea glandulosa	Mauritius	"
Purphiok	Tupistra sp.	Sikkim	"
Camomile flowers	Anthemis nobilis	Britain	Mixed with tobacco

(Table 19—continued)

Beet leaves	Beta vulgaris	France	Recommended as substitute
Akel		Algeria	Mixed with tobacco
Trouna		"	"
Kauw goed	Mesembryanthemum tortuosum	Cape	Chewed
Angelica root	Archangelica officinalis	Lapland	"
Monkey bread leaves	Adansonia digitata	W. Africa	Snuffed
Rhododendron leaves	Rhododendron campanulatum	India	Snuffed
Brown dust of petioles of	Kalmia sp. Rhododendron sp.	N. America	"
Asarabacca	Asarum Europceum	Europe	"
Grimstone's eye snuff	Various plants	Britain	"
Various indigenous plants		Erzegebirge	Mixed with snuff
Woodruff	Asperula odorata	Britain	Snuffed
Amadou ashes	Polyporus igniarius	Kamtschatka	

II. Opium

Smyrna opium	Papaver somniferum	Levant	Smoked, etc.
Constantinople do.	"	Turkey	"
Egyptian do.	"	Egypt	"
Trebizond do.	"	Persia	"
Bengal do.	"	India	"
Garden Patna do.	"	"	"
Malwa do.	"	"	"
Cutch do.	"	"	"
Kandeish do.	"	"	"

(Table 19—continued)

Vulgar name	Botanical name	Where used or cultivated	How used
English do.	"	"	"
French do.	"	"	"
German do.	"	"	"
Lactucarium	Lactuca sativa	Britain	Subs. for opium
"	" virosa	"	"
"	" scariola	"	"
"	" altissima	"	"
"	" sylvestris	"	"
"	" elongata	"	"
"	" taraxacifolia		
Dutchman's laudanum	Murucauja ocellata	Guiana	"
Ditto	" orbiculata	Jamaica	"
		Barbadoes	
Syrian rue seeds	Peganum harmala	Turkey	To produce intoxication
Seeds of	Sterculia alata	Silhet	Subs. for opium
Seeds of	Scopolia mutica	Arabia	To produce intoxication
Juice of	Chondrilla juncea	Lemnos	Subs. for opium
		III. Hemp	
Gunjah and bang	Cannabis indica	India, Africa	Smoked, etc.
Churrus (resin)	"	Nepaul, etc.	"
Powdered dacca and aloes	"	S.W. Africa	Snuffed

(Table 19—continued)

		IV. Betel	
Betel nuts	Areca catechu	Malay Penin	Chewed
"	Areca laxa	Andaman Is.	"
"	Areca Nagensis	E. Bengal	"
"	Areca Dicksoni	Malabar	"
Kassu (extract)	Areca catechu	India	"
Cowry (extract)	Areca catechu	Mysore	"
Kutt or catechu	Acacia catechu	India	"
Gambir	Uncaria gambir	Singapore, etc.	Chewed
"	Uncaria sp.		"
Betel pepper leaves	Chavica betel	Malay Penin.	Chewed with betel
"	Chavica siraboa	"	"
Blk. pepper leaves	Piper nigrum	Singapore	"
Ava pepper	Macropiper methysticum	S. Seas	"
Roots of	Derris pinnata	"	Subs. for betel
Guayabo bark	Psidium guayaba	Phillipines	"
Antipolo bark	"	"	"
		V. Coca	
Coca leaves	Erythroxylon coca	Peru	Masticatory

(*Table 19—continued*)

Vulgar name	Botanical name	Where used or cultivated	How used
		VI. Thorn-apple	
Florispondio seeds	Datura sanguinea	N. Granada	Drank in infusion
Thorn-apple leaves	" stramonium	Europe	Smoked
" " seeds	" arborea	Peru	"
" " "	" fatuosa	Egypt	"
" " "	" ferox	China	"
" " "	" tatula	Asia	By the Delphic oracle
Belladonna leaves	Atropa belladonna	Europe	Smoked
Henbane leaves	Hyoscyamus niger	India	Mixed with haschisch
Leaves of	Rhododendron chrysanthum	Siberia	Chewed
Flowers of	Rhododendron arboreum	India	"
Foxglove leaves	Digitalis prupurea	"	Mixed with haschisch
		VII. Amanita	
Fly agaric	Amanita muscaria	Siberia	Swallowed